THE BAOFENG RADIO BIBLE

12 BOOKS IN 1

The Ultimate Guerrilla's Guide To The Baofeng Radio. Stay Connected And Prepared For Any Emergencies, Wars & Natural Disasters.

By

Maxwell Cipher

TABLE OF CONTENT

INTRODUCTION 6

CHAPTER 1: THE BAOFENG RADIOS 12

   HISTORY AND EVOLUTION OF BAOFENG RADIOS................................ 13

   WHY BAOFENG STANDS OUT IN THE MARKET 15

   GENERAL LIST OF THE VARIOUS BAOFENG MODELS 17

   WHY UV-5R IS THE BEST MODEL TO USE IN THE USA 19

CHAPTER 2: ADVANCED FEATURES, HACKS, AND BEST PRACTICES............ 22

   HIDDEN FEATURES OF BAOFENG RADIOS 24

   EXPERT HACKS AND TIPS 26

   BEST PRACTICES FOR OPTIMAL USE 29

CHAPTER 3: PRACTICAL RADIO COMMUNICATION 32

   ADVANCED RADIO ETIQUETTE AND PROTOCOLS.............................. 33

   REAL-WORLD COMMUNICATION SCENARIOS: 34

   DECODING RADIO JARGON: A SIMPLIFIED GUIDE 36

   SPEAK THE LANGUAGE 39

   CHANNELS 41

   BASIC RADIO ETIQUETTE 49

CHAPTER 4: ADVANCED RADIO COMMUNICATION 51

   IMPORTANCE OF ADVANCED RADIO COMMUNICATION:...................... 52

   DATA HANDLING 55

   BENEFITS OF USING ADVANCED REPEATER SERVERS 57

CHAPTER 5: MASTERING RADIO PROGRAMMING 61

   A STEP-BY-STEP GUIDE TO MANUAL PROGRAMMING 64

   ADVANCED PROGRAMMING WITH CHIRP: TIPS AND TRICKS 66

   CUSTOMIZING CHANNELS AND FREQUENCIES FOR VARIOUS SCENARIOS 68

CHAPTER 6: SIGNAL ENHANCEMENT, RANGE, AND LIMITATIONS 71

TECHNIQUES TO BOOST SIGNAL RANGE 72

ADDRESSING RANGE MISCONCEPTIONS AND LIMITATIONS 74

SELECTING THE PERFECT ANTENNA 77

POWER SYSTEMS 79

POWER CONNECTORS 82

RADIO BAND GUIDE 83

CHAPTER 7: BAOFENG IN EMERGENCIES: A SURVIVAL GUIDE 85

CRAFTING AN EFFECTIVE EMERGENCY COMMUNICATION PLAN 88

ESSENTIAL FREQUENCIES FOR CRISIS SITUATIONS 90

REAL-LIFE TESTIMONIES: BAOFENG IN ACTION 93

CHAPTER 8: TROUBLESHOOTING, MAINTENANCE, AND BATTERY ISSUES 95

COMPREHENSIVE GUIDE TO COMMON ISSUES 98

ADDRESSING BATTERY PACK CONCERNS 100

PREVENTIVE MAINTENANCE TIPS: 104

DIAGRAM OF BAOFENG UV5R RADIO: 104

STORE BATTERIES PROPERLY 105

CLEAN BATTERIES AND TWO-WAY RADIOS REGULARLY 106

CHAPTER 9: NAVIGATING THE LEGAL LANDSCAPE 107

UNDERSTANDING FCC REGULATIONS 108

LICENSING DEMYSTIFIED: A STEP-BY-STEP GUIDE 111

ETHICAL CONSIDERATIONS FOR RADIO USERS 112

CHAPTER 10: BUILDING COMPREHENSIVE COMMUNICATION PLANS 115

DESIGNING EFFECTIVE COMMUNICATION STRATEGIES 118

UTILIZING BAOFENG RADIOS IN PLANS 121

CASE STUDIES AND EXPERT INSIGHTS 123

CHAPTER 11: EFFICIENT COMMUNICATION WITH TRIGRAMS 127

INTRODUCTION TO TRIGRAMS IN RADIO COMMUNICATION 128

COMPLETE TRIGRAM LIST FOR PRACTICAL USE 129

Tips and Best Practices for Using Trigrams 132

CHAPTER 12: DECODING RADIO SIGNALS AND MESSAGES 134

Importance of Decoding in Radio Communication 137

Comprehensive Decode List 138

Case Studies: Real-life Examples of Decoding Messages 140

CONCLUSION 142

GLOSSARY 144

THANK YOU FOR CHOOSING
'THE BAOFENG RADIO BIBLE: 12 BOOKS IN 1'
FROM MAXWELL CIPHER TO YOU, ADVENTUROUS READERS

Dear radio enthusiasts,

I am Maxwell Cipher, and I want to thank you for choosing **"The Baofeng Radio Bible"**. This book is written with a passion for adventure and communication, dedicated to you and all who are curious and brave enough to explore the world of radio communication.

Your decision to delve into this comprehensive guide reflects your commitment to staying connected and prepared in the face of emergencies, wars, and natural disasters. In these pages, I've poured extensive research, practical experience, and a deep passion for effective communication during critical times. Whether you're a seasoned radio enthusiast or a newcomer to the world of Baofeng radios, this book aims to equip you with all the knowledge and skills needed to master your device.

Before you immerse yourself in the secrets of Baofeng radios, let me share a review that particularly touched my heart:

"As an outdoor enthusiast who loves off-the-grid adventures, "The Baofeng Radio Bible: 12 Books in 1" was a fantastic find! This book isn't just a guide; it's a comprehensive masterclass on using Baofeng radios. The author's expertise shines through every page, making complex topics feel approachable. What really stood out to me was the practical approach to staying connected in emergencies. It's like having a survival expert guiding you through the nuances of radio communication. Whether it's a weekend camping trip or preparing for unforeseen events, this book has boosted my confidence in staying connected with my crew, no matter where we are. A must-read for anyone serious about reliable communication in any scenario!" - Cip

Unfortunately, many readers don't realize how challenging it is to receive feedback and how much they help authors like me to spread the word about our work. Now it's your turn to discover the book and, if these initial chapters inspire you, please share your experience by leaving an honest feedback on Amazon.

I would be very grateful if you could support me and help bring this book to more people. Never underestimate the power of a feedback; it's the support that makes a difference in informing and inspiring other readers.

Leaving your feedback is easy, and I appreciate every single one. Go to the ORDERS section of your Amazon account and click on the "Write a review for the product" button, or scan this QR code. It will automatically take you to the review section.

If, however, you think something in the book could be improved, I invite you to share your ideas with me directly at: **info@survivalhorizon.com**.

Your journey through **"The Ultimate Guerrilla's Guide To The Baofeng Radio"** is more than just a learning experience; it's a step towards ensuring safety and readiness in an unpredictable world. As you explore each chapter, I encourage you to practice, experiment, and embrace the art of radio communication.

For further updates, additional resources, and a platform to connect with fellow readers, please visit **www. survivalhorizon.com**

Stay safe and connected,
Let's start this journey together, *Maxwell Cipher*

DON'T MISS OUT – YOUR ADVENTURE CONTINUES ON PAGE 51!

Make your way there now and use your phone to scan the QR Code.

Awaiting you is a treasure trove of knowledge: 3 Exclusive Bonuses meticulously crafted to enhance your Baofeng Radio experience.

These bonuses are my way of saying thank you for embarking on this journey with me. So, take the next step, unlock these special gifts, and dive deeper into the world of Baofeng Radio mastery!

INTRODUCTION

The Baofeng phenomenon typically alludes to the popularity and widespread use of Baofeng two-way radios, manufactured by the Chinese company Fujian Nanan Baofeng Electronics Co., Ltd. These radios have gained immense popularity due to their affordability, ease of use, and functionality. Below, I outline a few key points encapsulating the Baofeng phenomenon:

Affordability:

- **Low Cost:** Baofeneh ma Crig radios are notably inexpensive, making them accessible to a broad audience including hobbyists, outdoor enthusiasts, and professional users.
- **Cost-effective Communication Solution**: For individuals or small businesses requiring a reliable communication device without a significant investment, Baofeng offers a viable option.

Accessibility:

- **Availability**: The radios are widely available for purchase online, providing easy accessibility to users worldwide.
- **Ease of Use**: Baofeng devices are known for their user-friendly nature, with straightforward controls and interfaces that are approachable for newcomers.

Functionality:

- **Versatile Features**: Despite their low cost, Baofeng radios boast various features, often matching the capabilities of more expensive models from established brands in the industry.
- **Frequency Range**: The radios can operate on different frequency bands, making them suitable for various communication purposes including amateur radio (ham radio) and FRS/GMRS.

Popularity Among Specific User Groups:

- **Outdoor Enthusiasts**: Individuals engaged in hiking, camping, and other outdoor activities often use Baofeng radios for communication, given the devices' robustness and reliability.
- **Amateur Radio Operators**: The ham radio community appreciates the affordability and functionality of Baofeng radios, especially those entering the hobby and looking for cost-effective equipment.
- **Emergency Preparedness Community**: Individuals focused on preparing for emergencies or disasters often include Baofeng radios in their kits due to their ease of use and versatility.

Controversies:

- **Legal Concerns**: In some jurisdictions, legal issues or concerns arise related to the use of Baofeng radios, particularly regarding their transmission capabilities on certain frequencies reserved for licensed services. Users must be aware of and comply with their local regulations and laws governing two-way radio use.

Impact on the Market:

- **Competitive Pressure**: The affordability and functionality of Baofeng radios have exerted competitive pressure on other brands in the two-way radio market, often leading to price reductions and innovation.

Limitations:

- **Quality Variance**: While Baofeng radios are functional, there might be a quality variance between units, and they may not match the durability and performance of higher-end models from established brands.

Downsides:

1. **Quality Inconsistencies:**

Due to their low price point, the build quality and durability of Baofeng radios can be inconsistent. Some units may work flawlessly, while others might experience technical issues or malfunction over time.

2. **Limited Support and Warranty:**

Baofeng radios often come with limited customer support and warranties compared to more established brands. Users might find it challenging to get assistance when encountering issues with their devices.

3. **Regulatory Compliance:**

In some countries, strict regulations regarding the use of two-way radios exist, and some Baofeng models may not comply with all of these. Users need to ensure they are using the radios legally to avoid fines or penalties.

4. **Learning Curve for New Users:**

While generally user-friendly, Baofeng radios might still present a learning curve for individuals new to two-way radios. Mastering the device's functions and features may take some time and patience.

5. **Frequency Accuracy and Stability:**

Baofeng radios may not be as accurate or stable in maintaining specific frequencies compared to higher-end models, which can be crucial for certain applications where precision is key.

6. **Battery Life:**

The battery life of Baofeng radios might not be on par with more expensive models. Depending on usage, users might need to recharge or replace batteries more frequently.

Things to Consider:

1. **Legal Restrictions:**

Be aware of and understand the legal restrictions and licensing requirements for using two-way radios in your jurisdiction.

2. **Antenna Upgrades:**

The stock antennas that come with Baofeng radios are often considered subpar by users. Investing in a better-quality antenna can significantly improve the radio's performance.

3. **Programming Software:**

While the radios can be programmed manually, using programming software might simplify the process for users. However, the quality and user interface of the software can vary.

4. **Interference Issues:**

Baofeng radios might experience interference issues in certain environments, affecting the clarity and reliability of communication.

5. **Compatibility:**

Before purchasing, check the compatibility of Baofeng radios with other brands and models to ensure seamless communication if you are planning to use them in a mixed fleet.

6. **Intended Use:**

Consider your primary purpose for the radio. If you require a device for critical communication or professional use, investing in a higher-end model might be more prudent.

CHAPTER 1: THE BAOFENG RADIOS

Baofeng radios, hailed as versatile handheld transceivers, have become synonymous with efficient and affordable communication across a diverse range of frequency bands. These remarkable devices are the creation of the esteemed Chinese company, Fujian Nanan Baofeng Electronic Co. Ltd., which commenced its journey in 2001 in the picturesque region of Fujian, China.

The remarkable success and enduring legacy of Baofeng radios can be attributed to their adaptability and widespread appeal. These portable communication devices, renowned for their multifaceted capabilities, have found a dedicated following among hobbyists, professionals, and emergency responders alike.

Fujian Nanan Baofeng Electronic Co. Ltd. has, since its inception, demonstrated a remarkable commitment to crafting high-quality two-way radios and complementary accessories. The company's dedication to producing reliable and feature-rich communication tools has earned them recognition and admiration from a diverse spectrum of users.

As communication technology evolved over the years, Baofeng radios kept pace with the changing landscape. They have consistently been at the forefront of innovation, introducing an array of models that cater to a wide spectrum of user requirements. These radios are adept at operating across the VHF and UHF frequency bands, offering dual-band functionality that allows users to communicate seamlessly over extended distances and in diverse environments.

Moreover, Baofeng radios have also established themselves as budget-friendly options without compromising on functionality. They boast a plethora of channels, user-friendly programmable settings, clear LCD displays, and long-lasting battery life, which make them indispensable tools for various applications, including outdoor adventures, community-based communication networks, and public safety initiatives.

In particular, Baofeng radios have emerged as pivotal components in emergency communication and disaster relief efforts. Amateur radio operators, who frequently volunteer their services during crises, rely on these devices for their reliability and extensive features, thereby further cementing Baofeng's position as a trusted name in the world of two-way communication.

HISTORY AND EVOLUTION OF BAOFENG RADIOS

Ever since they were first introduced to the market, Baofeng radios, which are manufactured by the Chinese business Fujian Nan'an Baofeng Electronics Co., Ltd., have left a considerable imprint on the world of two-way communication. These multipurpose portable radios have a long and illustrious history, and during the course of their development, they have progressed significantly, becoming a vital instrument for radio aficionados, professionals, and first responders.

The history of Baofeng radios can be traced back to the late 20th century when the company was founded in 2001. Initially, they focused on producing land mobile radios, but their breakthrough came in 2005 when they introduced their first amateur radio, the UV-3R. This compact and affordable handheld transceiver quickly gained popularity among amateur radio operators, also known as "ham" radio enthusiasts.

Over the years, Baofeng radios have evolved in terms of technology, features, and design. They've introduced various models, each catering to specific needs. These radios typically operate in the VHF and UHF frequency ranges and offer dual-band capabilities, allowing users to communicate over a wide range.

One of the key factors behind Baofeng's success is their commitment to affordability without compromising on functionality. Their radios often feature a multitude of channels, programmable settings, LCD displays, and long-lasting batteries, making them suitable for various applications, including outdoor adventures, community-based communications, and public safety.

Furthermore, Baofeng radios have played a vital role in emergency communication and disaster relief efforts, thanks to their reliability and widespread use among amateur radio operators who often volunteer their services during crises.

The history and evolution of Baofeng radios serve as a testament to their enduring relevance in the realm of two-way communication, showcasing the brand's dedication to innovation and affordability in an ever-evolving industry.

Evolution of Baofeng Radios:

1. **Early Years (2001-2010):**
- **Founding**: The company was established with an emphasis on providing affordable, compact, and reliable communication devices.
- **Initial Products**: Baofeng's early offerings included basic two-way radios designed for short-range communication.

2. **Rapid Development (2010-2015):**

- **UV-5R Series Release**: Baofeng's UV-5R became one of its most popular models during this period due to its affordability, durability, and versatility. The UV-5R operates on VHF and UHF bands, allowing users to communicate over both short and long distances.

- **Growing Market Share**: With the success of the UV-5R and similar models, Baofeng gained a significant market presence in the two-way radio sector.

- **Global Expansion**: The brand expanded its market beyond China, reaching users in North America, Europe, and other regions.

3. **Technological Improvements (2015-2020):**

- **Digital Radios**: Baofeng introduced digital radios in addition to its traditional analog models, catering to the needs of users seeking clearer communication and advanced features.

- **Feature Enhancement**: The radios incorporated additional features like FM radio, LED flashlight, and emergency alarms, making them more appealing to various user groups.

  - **Improved Battery Life**: The company focused on enhancing the battery life of its devices, ensuring longer usage times for users.

4. **Recent Developments (2020-present):**

- **Diversification**: Baofeng began offering diverse models with various features to cater to different market segments, from casual users to professionals in the field.

- **Software Improvements**: The company worked on improving the programming software for its radios, allowing for more straightforward customization and channel programming.

Factors Contributing to Baofeng's Success:

- **Affordability**: Baofeng radios are known for being cost-effective, providing good value for the price, which attracted many users, particularly amateurs and enthusiasts.

- **Versatility**: The radios are versatile, functioning in various environments and situations, making them suitable for different users, including hikers, campers, and event coordinators.

- **Community Support**: The brand has fostered a community of users who share insights, tips, and programming guides, contributing to a supportive ecosystem for Baofeng radio owners.

Challenges and Controversies:

Regulatory Issues: Baofeng radios have faced challenges with regulatory compliance in some regions. For example, in the United States, the Federal Communications Commission (FCC) has raised concerns about the radios' compatibility with federal regulations, leading to scrutiny and adjustments in the devices' distribution and usage.

Quality Consistency: While Baofeng radios are affordable, there have been concerns about the consistency of quality and durability across different units and models.

WHY BAOFENG STANDS OUT IN THE MARKET

Baofeng radios have not only established themselves in the competitive two-way radio market, but have also carved out a distinctive niche for several compelling reasons. Their standout presence can be attributed to a confluence of factors that set them apart from the competition, making them a preferred choice for a diverse range of users.

First and foremost, affordability plays a pivotal role in their popularity. Baofeng radios have managed to strike a harmonious balance between cost and functionality, offering a cost-effective solution without compromising on essential features. This affordability makes them accessible to a wide spectrum of users, from amateur radio enthusiasts on a budget to professionals in need of dependable communication tools.

Moreover, the flexibility and adaptability of Baofeng radios are noteworthy. Their extensive lineup features various models, each tailored to meet specific needs. Whether it's a compact handheld transceiver for outdoor adventurers, a high-power unit for long-range communication, or a versatile model for public safety personnel, Baofeng has a radio to suit the purpose.

Another key strength is their dual-band capability, allowing users to communicate on both VHF and UHF frequencies. This dual-band feature significantly enhances their usability, enabling users to access a wider range of channels and networks, whether it's for recreational use or critical emergency communications.

Baofeng radios also stand out for their versatility in programming and customization. They often come equipped with a plethora of channels, a user-friendly interface, and LCD displays for convenient operation. This adaptability makes them suitable for various applications, from local community-based communications to complex public safety operations.

Furthermore, Baofeng radios have played a crucial role in the realm of emergency communication and disaster relief. Their reliability and wide adoption among amateur radio operators make them a valuable asset during times of crises, as they enable volunteers to provide vital communication services in the absence of traditional infrastructure.

In essence, the standout presence of Baofeng radios is the result of their unwavering commitment to innovation, affordability, and functionality. Their ability to cater to the diverse needs of users across different sectors is a testament to their enduring relevance and growing importance in the world of two-way communication.

Below are some of the key factors that have contributed to Baofeng's standout presence in detail:

1. **Affordability:**
    -   **Budget-Friendly**: Baofeng radios are remarkably affordable compared to devices from other brands offering similar features.
    -   **Cost-Effective Communication**: For many users, particularly amateurs and hobbyists, Baofeng provides a cost-effective entry point into the world of two-way radios without the need for a significant investment.

2. **Versatility:**
    -   **Wide Range of Frequencies**: The radios cover a broad spectrum of frequencies, allowing users to communicate across various bands.
    -   **Multiple Functions**: With features like FM radios, LED flashlights, and emergency alarms, Baofeng radios serve multiple purposes beyond communication.

3. **Ease of Use:**
    -   **User-Friendly Interface**: Baofeng devices typically have interfaces that are straightforward and easy for beginners to navigate.
    -   **Easy Programming**: With accessible programming options, users can easily set up and customize their devices to suit their communication needs.

4. **Global Community:**
    -   **Supportive User Base:** The brand has cultivated a loyal and active community of users worldwide who share tips, programming guides, and support.
    -   **Online Forums and Groups:** There are numerous online platforms where Baofeng users gather to exchange knowledge and offer assistance to one another, fostering a sense of camaraderie and support among users.

16

5. **Wide Product Range:**
   - **Variety of Models:** Baofeng offers a diverse range of models catering to different user requirements, from basic to more advanced devices.
   - **Upgrades and Innovations:** The company regularly updates its product line, introducing new models and features to meet the evolving needs of its customer base.

6. **Global Accessibility:**
   - **Worldwide Availability:** Baofeng radios are available for purchase in many countries, making them accessible to a global customer base.
   - **International Distribution Networks:** The brand has established distribution channels that make it easy for customers around the world to acquire their products.

7. **Durable and Compact Design:**
   - **Robust Build:** Many Baofeng radios are known for their durable construction, capable of withstanding rough handling and challenging environmental conditions.
   - **Portable:** The radios are often compact and lightweight, making them convenient for users on the move.

Challenges to Consider:

While Baofeng radios stand out for various positive reasons, potential users should also be aware of challenges, like:

- **Regulatory Compliance**: In certain regions, there may be regulatory restrictions on the use of Baofeng radios due to concerns about interference with licensed communication services.
- **Quality Variations**: Due to their low cost, some users have reported variations in the quality and durability of Baofeng radios, which may impact their long-term reliability.

GENERAL LIST OF THE VARIOUS BAOFENG MODELS

Baofeng offers a wide range of two-way radios, with various models designed to cater to different users. While the company continues to release new products, below is a general list of some of the popular models up until my knowledge cut-off in 2022:

UV Series:

1. **UV-5R**: This model is among the most popular and widely recognized Baofeng radios. It operates on both VHF and UHF bands and features a dual-band display, dual-frequency display, and dual-standby functions. It is known for its affordability and versatility.

2. **UV-82**: An upgraded version of the UV-5R, the UV-82 boasts a more ergonomic design, a louder speaker, and a new chipset for enhanced reception. It also operates on dual bands.

3. **UV-9R**: The UV-9R is a more rugged and waterproof model designed for more strenuous outdoor activities and professional use. It also features improvements in battery life and antenna design.

BF Series:

1. **BF-F8HP**: This model is considered a third-generation UV-5R with more power, a hardened durable radio shell, and a new chipset and PCB board that outperform the range, accuracy, and output of previous Baofeng models.

2. **BF-888S**: The BF-888S is a straightforward, easy-to-use, and incredibly affordable model designed for short-range communication. It operates on the UHF band and is often used in settings like restaurants, hotels, and other on-site businesses.

DM Series:

1. **DM-1701**: The DM-1701 is a dual-band, analog and digital DMR radio with a colorful screen and GPS function. It's designed to meet the requirements of both amateur radio enthusiasts and professional users.

2. **DM-5R**: This is a budget-friendly DMR radio model that supports both traditional analog and digital modes. It's often used by amateurs who are transitioning from analog to digital communications.

GT Series:

1. **GT-3**: The GT-3 is an upgraded version of the UV-5R with a new generation chipset, enhanced features, and a more rugged design. It also comes with a better antenna that offers more gain.

Other Models:

1. **UV-3R**: A compact mini radio with dual-band functionality.

2. **UV-6R**: An improved and more water-resistant version of the UV-5R.

3. **UV-B6**: Offers features similar to the UV-5R but with a larger battery and a different form factor.

WHY UV-5R IS THE BEST MODEL TO USE IN THE USA

The Baofeng UV-5R has earned a well-deserved reputation as one of the most highly sought-after and beloved models among a diverse user base in the United States. Radio enthusiasts, preppers, outdoor adventurers, emergency responders, and a wide range of users embrace the UV-5R for its exceptional performance and versatility in the realm of short-range communication devices.

Radio enthusiasts, also known as amateur radio operators or "ham" radio enthusiasts, have long appreciated the Baofeng UV-5R for its outstanding features. With dual-band capabilities that operate on both VHF (Very High Frequency) and UHF (Ultra High Frequency), the UV-5R allows enthusiasts to explore a wide spectrum of frequencies and engage in long-distance communications. The ability to program a plethora of channels and customize settings to suit individual preferences makes this model particularly appealing to this dedicated group of radio aficionados.

Preppers, individuals who prepare for potential emergencies and disasters, also embrace the UV-5R as an essential tool in their preparedness kits. These radios provide a lifeline during crises when traditional communication infrastructure may be compromised. The UV-5R's rugged build and long-lasting battery are well-suited for the unpredictable challenges that preppers may face.

Outdoor enthusiasts, such as hikers, campers, and boaters, find the Baofeng UV-5R to be a reliable companion on their adventures. Its compact size and lightweight design make it easy to carry, and the radios often feature built-in flashlight functions, enhancing their utility in the great outdoors.

In addition, emergency responders and community-based organizations have adopted the UV-5R as a cost-effective means of maintaining communication networks during localized emergencies and public events. The UV-5R's affordability and versatility have made it a popular choice for these critical roles.

In the US, the Baofeng UV-5R is a popular choice for dependable, short-range communication solutions, regardless of user demands.

Below are the reasons why the UV-5R is considered by many as an excellent choice:

Affordability:

- Cost-Effective: The UV-5R is priced very affordably, making it an accessible entry point for individuals new to two-way radios or those needing multiple units without a significant investment.

Versatility:

- Dual-Band: As it operates on both the VHF and the UHF bands, the UV-5R gives users access to a broad range of frequencies to accommodate a variety of communication requirements.
- Features: It comes with various features like an FM radio, a built-in flashlight, and an emergency alarm, making it practical for different situations.

Ease of Use:

- User-Friendly: With a straightforward interface and an easily navigable menu, the UV-5R is simple for beginners to use and program.
- Programming Software: Users can take advantage of available software to program the device via a computer, simplifying the process of setting up channels and frequencies.

Community Support:

- Wide User Base: The UV-5R has a large community of users who actively share tips, guides, and programming information, providing support to newcomers.
- Online Resources: Numerous forums, social media groups, and websites are dedicated to Baofeng UV-5R users, offering a wealth of knowledge and assistance.

Durability:

- Robust Design: The device is known for its sturdy construction and durability, capable of withstanding rough handling to an extent.
- Portability: Its compact and lightweight design makes it easy to carry, suited for outdoor, recreational, and emergency use.

Accessory Availability:

- Wide Range of Accessories: There's a wide market of compatible accessories available for the UV-5R, including antennas, batteries, earpieces, and programming cables.
- Customization: Users can easily enhance and personalize their devices to fit specific needs and preferences through available accessories.

Firmware Upgrades:

- Continuous Improvement: Over the years, the UV-5R has received various firmware upgrades and updates that have improved its performance and reliability.

Considerations and Limitations:

- Regulatory Compliance: Users must ensure they comply with FCC regulations when operating the UV-5R, as improper use on unauthorized frequencies can lead to fines and legal consequences.

- Licensing Requirements: To operate on certain bands and frequencies, users might need to obtain an amateur radio license from the FCC.

- Learning Curve: While user-friendly, new users might still face a learning curve in programming and operating the device efficiently.

CHAPTER 2: ADVANCED FEATURES, HACKS, AND BEST PRACTICES

Baofeng radios have gained significant popularity among amateur radio enthusiasts due to their affordability and widespread use as portable two-way radios. The Baofeng UV-5R is well recognized as a highly sought-after model, yet it is important to note that there are other alternative options, each distinguished by its own array of characteristics. The following are a selection of advanced functionalities, strategies, and recommended methods that may be used to optimize the performance and utility of your Baofeng radio.

Advanced Features:

Baofeng radios often come equipped with an impressive array of advanced features. These can include dual-band operation, allowing you to transmit and receive on both VHF and UHF frequencies. This capability opens up a broader spectrum of communication options, making the radios suitable for a wide range of activities, from amateur radio conversations to monitoring public service frequencies. Some models offer cross-band repeat functionality, enabling the retransmission of signals between different frequency bands, further enhancing their versatility.

Dual Watch Function:

Baofeng radios often have a Dual Watch function that allows monitoring of two frequencies. This is useful for listening to two different channels without needing to switch back and forth manually.

VOX Function:

VOX (Voice Operated Exchange) allows hands-free communication. The radio transmits when it detects your voice without the need to press any button.

FM Radio Receiver:

Some Baofeng radios can receive commercial FM radio broadcasts, allowing you to listen to music, news, and more.

Hacks:

Extended Frequency Range:

With specific software (like CHIRP), you can expand the frequency range of your Baofeng radio. However, transmitting on unauthorized frequencies is illegal in many jurisdictions, so proceed with caution.

Programming via Computer:

Utilizing a computer for programming radio stations offers a notably more convenient approach compared to the manual method. One option for programming a Baofeng radio is to use freely available software such as CHIRP.

Custom Firmware:

Some enthusiasts develop custom firmware that can unlock additional features or improve the performance of the radio. Be cautious as installing unofficial firmware can void your warranty and potentially brick your device.

Best Practices:

To make the most of your Baofeng radio, it's essential to adhere to best practices for effective and safe operation. This includes proper programming of channels and frequencies, understanding the nuances of repeater systems, and adopting efficient communication techniques. Additionally, being well-versed in your local radio licensing and usage regulations is vital to ensure you're using your Baofeng radio responsibly and within the bounds of the law.

Antenna Upgrade:

The stock antenna may not offer the best range or reception. Consider upgrading to a higher-quality aftermarket antenna to improve performance.

Power Conservation:

If you want to conserve power, use the low power setting unless you need to communicate over longer distances.

Respect Legal Restrictions:

In many countries, you need a license to operate on certain frequencies. Always respect these restrictions to avoid fines or other legal issues.

Emergency Use:

Having a radio in case of an emergency is a good practice. Familiarize yourself with the emergency channels in your area and know how to use your radio to call for help if necessary.

Protect Your Gear:

Consider investing in protective cases and perhaps a waterproof bag if you plan to use your Baofeng radio outdoors in various conditions.

HIDDEN FEATURES OF BAOFENG RADIOS

Baofeng radios, namely the Baofeng UV-5R model, have garnered significant recognition within the realm of two-way radios owing to their cost-effectiveness and multifunctionality. Although these radios are extensively used by both amateur radio enthusiasts and professionals, there exist several concealed capabilities and techniques that may not be commonly known to many users. This article aims to examine several elements of Baofeng radios that are not well-recognized, which have the potential to improve one's radio experience.

1. **Frequency Range Expansion:** Baofeng radios possess a restricted frequency range; nevertheless, a hidden functionality exists that enables the expansion of this range. By inputting a certain sequence of keystrokes, it becomes feasible to access a more extensive range of frequencies, hence enabling the reception of a greater selection of channels. However, it is important to verify local restrictions before using this functionality, since some frequencies could be allocated for exclusive usage by certain users.

2. **CTCSS and DCS Tones**: Baofeng radios support Continuous Tone-Coded Squelch System (CTCSS) and Digital-Coded Squelch (DCS) tones. These features help you filter out unwanted transmissions and interference. The hidden feature here is the ability to set different CTCSS or DCS tones for both transmit and receive, giving you more control over who you can communicate with.

3. **Dual Watch and Dual Reception**: Baofeng radios allow you to monitor two frequencies simultaneously, which is a great feature for monitoring two channels at once. However, there's a hidden feature that enables dual reception, meaning you can actively receive transmissions on two frequencies simultaneously. This can be especially useful for search and rescue operations or monitoring multiple channels during an event.

4. **Reverse Frequency Function**: If you need to quickly switch between transmit and receive frequencies, the Baofeng radio offers a reverse frequency function. By pressing a specific key combination, you can switch the transmit and receive frequencies without having to manually reprogram the radio.

5. **Customizable Channel Names**: Baofeng radios allow you to assign custom names to your channels for easier identification. Many users are unaware of this feature, which can be incredibly helpful, especially when dealing with numerous channels. You can assign names like "Family," "Emergency," or "Work" to make channel selection more intuitive.

6. **VOX (Voice-Operated Transmission):** Baofeng radios support VOX mode, allowing you to transmit without pressing the Push-To-Talk (PTT) button. This feature is useful when your hands are occupied, but the hidden trick is that you can adjust the sensitivity level of VOX to suit your specific needs.

7. **Scan Mode Options**: While scanning channels is a standard feature, there are hidden scan mode options available. You can set the scan to pause on active channels, skip unwanted frequencies, or even change the scan direction (up or down) to customize your scanning experience.

8. **Battery Saver Mode**: To conserve battery life, Baofeng radios have a battery saver mode. Many users don't realize that this mode can be customized to activate after a certain period of inactivity, making it even more efficient.

9. **Emergency Alerts:** Baofeng radios can transmit emergency signals that are often overlooked. You can send emergency alerts to other radios by pressing a specific key combination. This feature can be a lifesaver in critical situations.

10. **Adjustable Transmit Power**: Baofeng radios come with multiple power settings. The hidden feature here is the ability to manually adjust your transmit power, allowing you to save battery when necessary or increase your range in more challenging environments.

11. **Programming via Computer:** While not exactly a hidden feature, many users are unaware that you can program your Baofeng radio via a computer. There are software and programming cables available that simplify the process of setting up your radio and customizing frequencies and settings.

12. **Silent Operation:** For discreet communication, Baofeng radios can be set to silent operation, which mutes all beeps and alerts. This is an excellent feature for hunting, security, or any situation where noise could be a concern.

13. **Timer Functions**: Baofeng radios include timer functions, which can be used for various purposes, such as setting countdown timers for specific tasks or configuring the radio to turn off after a certain period of inactivity.

In conclusion, Baofeng radios offer a wide array of features and hidden functions that can greatly enhance your radio communication experience. These features range from simple convenience options like customizable channel names to more advanced capabilities like frequency range expansion and dual reception. By exploring and utilizing these hidden features, you can make the most of your Baofeng radio and adapt it to your specific needs, whether you are an amateur radio operator, outdoor enthusiast, or a professional in need of reliable communication tools. Always remember to comply with local regulations and ethical guidelines when using these features to ensure responsible and effective radio communication.

EXPERT HACKS AND TIPS

When using Baofeng radios, there are several advanced techniques and strategies that may be used to enhance performance and maximize the overall user experience. Below are a few recommendations:

Expert Hacks for Maximizing Your Baofeng Radio Experience:

While Baofeng radios offer a wide range of features right out of the box, there are several expert-level hacks and adjustments that can take your radio experience to the next level. These tweaks and modifications are typically geared towards seasoned users who want to unlock hidden potential and fine-tune their devices for specific purposes. However, it's essential to approach these hacks with caution and a clear understanding of their implications.

1. Firmware Modification: Pushing the Boundaries

One of the more advanced techniques in the Baofeng radio world involves modifying the device's firmware. This process can unlock hidden features or enhance performance, but it comes with some significant caveats. Firmware modification might void your warranty, be technically complex, and even run afoul of legal restrictions in some cases. Nevertheless, for users willing to take the risk and invest time in research and execution, it can be a rewarding endeavor.

2. Squelch Setting Adjustments: Clarity Amidst Noise

Adjusting the squelch settings on your Baofeng radio is a valuable technique for improving the clarity of transmissions. Squelch essentially helps filter out background noise, but finding the right balance is crucial. Setting it too high can cause you to miss weaker signals, while setting it too low can lead to constant interference. Expert users fine-tune their squelch settings to strike the ideal balance, ensuring a clean and clear audio experience.

3. Use of External Antennas: Expanding Your Reach

For users seeking an extended communication range, connecting your Baofeng radio to an external antenna is a game-changer. This enhancement may necessitate the use of adapters, but the payoff is a significantly stronger signal and an expanded reach. Whether you're in a remote outdoor location or need better signal penetration in urban environments, an external antenna can make all the difference.

4. Memory Channel Programming: Meticulous Organization

Expert users often employ specialized software to meticulously program memory channels. This method allows for the organized arrangement of frequencies for quick access to frequently used channels. Whether you're an amateur radio operator, a professional in the field, or a hobbyist with a vast range of interests, this approach saves time and ensures that your most essential channels are readily accessible.

5. Cross-Band Repeat Functionality: Extending Communication

Some Baofeng radios have the hidden capability to perform cross-band repeat functions, allowing them to act as a relay station. In situations where you need to extend your communication range, this feature can be invaluable. By setting up your radio as a relay, you can cover larger distances or even transmit from areas with poor signal reception.

6. Advanced Encryption: Secure Communication

For users who require secure and encrypted communication, there are ways to implement advanced encryption on Baofeng radios. This is particularly important for professionals in security, law enforcement, or any field where sensitive information needs to be protected. This feature provides an additional layer of security and privacy during communications.

7. Customized Power Levels: Battery Conservation

Baofeng radios come with multiple power settings, but expert users can further customize these levels to conserve battery life or boost signal strength when necessary. Tailoring the transmit power to your specific requirements can be critical, especially in situations where extended battery life is essential.

8. Advanced Scanning Techniques: Efficient Monitoring

Baofeng radios have scanning capabilities, but expert users take advantage of advanced scanning techniques to efficiently monitor a range of channels. This includes setting scan delays, defining scanning sequences, and configuring the radio to skip channels that are not relevant. These techniques ensure that you stay informed without unnecessary interruptions.

In summary, expert hacks for Baofeng radios can elevate your communication experience to new heights. However, these techniques should be approached with caution, especially regarding firmware modifications, as they can void warranties or raise legal issues. Users should always prioritize safety, legality, and responsible use when exploring these advanced features. Whether you're looking to extend your range, improve audio clarity, or enhance the functionality of your Baofeng radio, these expert-level adjustments offer the potential for a superior radio experience tailored to your specific needs.

Expert Tips:

Baofeng radios have gained recognition for their multifunctionality and cost-effectiveness, making them a favored option among many user demographics, including both amateur radio enthusiasts and professionals alike. To optimize the use of your Baofeng radio, we have curated a collection of specialized recommendations from experts that transcend basic functionalities. These suggestions aim to optimize your radio experience and cultivate your proficiency as an operator.

1. **Battery Management**: Battery management is a crucial element when using a Baofeng radio. To mitigate the risk of power depletion at critical moments, it is advisable to acquire additional batteries or contemplate the acquisition of a battery with enhanced capacity. The additional power source might serve as a crucial resource for prolonged excursions or critical circumstances. Furthermore, it is advisable to deactivate the radio when it is not being used to preserve the battery's lifespan.

2. **Hands-Free Kit**: A hands-free kit may significantly enhance the experience of those often on the move. The standard package generally consists of a superior-quality headset and microphone, enabling communication without the need to physically hold the radio device. Whether engaging in outdoor activities such as hiking and bicycling or undertaking task-oriented endeavors, the utilization of this particular device may substantially augment both convenience and safety.

3. **Understanding Frequency Bands**: To use your Baofeng radio effectively, it's essential to understand the frequency bands it can access and the legal restrictions associated with them. Familiarize yourself with which bands are designated for amateur use, commercial use, and emergency services. This knowledge ensures that you operate your radio within the legal framework and do not interfere with critical communications.

4. **RF Gain Control (If Available)**: Some Baofeng models come with an RF gain control feature, which allows you to adjust the radio's receiver sensitivity. This feature can be incredibly useful in situations with strong signals that might cause interference. By fine-tuning the RF gain, you can improve reception and maintain clear communication, even in challenging environments.

5. **Regular Firmware and Software Updates**: Regular firmware and software updates are advantageous for Baofeng radios. Ensuring regular updating of your radio's firmware is crucial for a multitude of reasons. Software updates have the capacity to address software flaws, improve overall performance, and perhaps enable additional functionalities. Keeping up to date with these updates guarantees that your radio operates at its highest level of efficiency and dependability. Please refer to Baofeng's official website for the latest updates and detailed information about their installation process.

6. **Practice Makes Perfect**: Like any tool, the more you use your Baofeng radio, the more proficient you become. Practice is key to mastering its features and quirks. Experiment with it in various situations, including outdoor adventures, emergency drills, or casual conversations. The more you use it, the more comfortable you'll become, making you a more effective operator.

By following these expert tips, you can unlock the full potential of your Baofeng radio. Battery management, the use of hands-free kits, understanding frequency bands and legal restrictions, adjusting RF gain, staying updated with firmware and software, and regular practice will not only enhance your radio experience but also ensure you're well-prepared for any situation in which reliable communication is vital. Whether you're a seasoned radio operator or new to the world of two-way radios, these tips will help you make the most of your Baofeng device.

Legal Considerations:

It is important to consistently adhere to the legal parameters established by the regulatory entities in one's jurisdiction. Engaging in the transmission of signals on certain frequencies without proper authorization, using excessive power levels, or utilizing modified equipment might result in financial penalties, seizure of equipment, or other legal consequences.

Safety Considerations:

When modifying or hacking radio equipment, there's always the risk of damaging the device or creating a safety hazard. Always follow safety guidelines, understand the risks involved, and if in doubt, consult with experienced radio operators or professionals in the field.

BEST PRACTICES FOR OPTIMAL USE

Baofeng radios offer a wide array of features that make them popular among amateur radio enthusiasts and professionals alike. To optimize the use of these radios, consider the following best practices:

1. **Understanding the Radio:**
   - Familiarize yourself with the functions and buttons of the radio.
   - Understand how to switch between frequencies, adjust volume, change channels, and use other essential features.

2. **Antenna Upgrade:**

   • The default antennas are usually not the best. Upgrade to a better antenna to improve signal reception and transmission range.

   • For handheld models like the UV-5R, Nagoya antennas are popular choices.

3. **Programming with Software:**

   • Use software like CHIRP to simplify the process of programming frequencies and channels.

   • This allows for more organized and efficient access to the channels you use most frequently.

4. **Power Settings:**

   • Be mindful of power settings. Using high power will drain the battery faster but increase range; low power conserves battery but decreases range.

   • Adjust according to your specific needs and situation.

5. **Battery Management:**

   • Keep spare batteries on hand, especially if you plan to use the radio for extended periods.

   • Charge batteries fully before use, and consider investing in higher-capacity batteries for longer life.

6. **Legal Compliance:**

   • Understand and comply with the legal requirements in your jurisdiction regarding radio frequency use.

   • Acquiring an amateur radio license is essential if you intend to transmit on ham bands.

7. **Emergency Preparedness:**

   • Learn how to access emergency frequencies and repeaters in your area.

   • Keep a list of important frequencies and codes, and understand how to communicate in emergencies.

8. **Hands-free Operation:**

   • Consider getting accessories like earpieces, microphones, or shoulder mics for hands-free operation.

   Using Squelch:

   • Properly set the squelch level to minimize background noise while not missing out on weaker transmissions.

9. **Device Protection:**

   • Invest in protective cases and screen protectors to safeguard your device from falls, impacts, and the elements.

10. Regular Maintenance and Updates:

• Keep the firmware updated, clean the device regularly, and check for any wear or damage to ensure optimal performance.

11. Practice and Training:

• Practice using the radio, join amateur radio clubs, participate in nets, and engage with the community to improve your skills and knowledge.

12. Safety First:

• Never transmit on frequencies reserved for emergency services or other authorized users, and always use the device responsibly and safely.

CHAPTER 3: PRACTICAL RADIO COMMUNICATION

In a world increasingly interconnected by the marvels of modern technology, radio communication stands as a timeless and indispensable medium that has shaped our lives in ways we often take for granted. From the humble walkie-talkie to the complex infrastructure supporting cellular networks, radio waves remain a vital part of our daily existence, fostering connectivity across vast distances. Practical radio communication, in all its forms, has woven a web of invisible threads that seamlessly link people, places, and information.

The origins of radio communication can be historically attributed to the latter half of the 19th century, during which individuals such as Guglielmo Marconi and Nikola Tesla played significant roles in laying the foundation for the subsequent wireless revolution. Over time, technological advancements have progressed at a remarkable rate, propelling us from the initial stages of Morse code communication to the contemporary era characterized by rapid data transfer. The advancement of radio technology has facilitated its widespread integration into many aspects of human existence, including the use of cellphones, automobile radios, public safety communication systems, and satellite connections that permit worldwide communication.

One of the defining features of practical radio communication is its ability to transcend physical barriers. Radio waves can effortlessly traverse obstacles that would obstruct other forms of communication, making them ideal for scenarios such as emergency response and disaster recovery. Whether it's a firefighter coordinating efforts inside a blazing building or an astronaut communicating with mission control from the far reaches of space, radio communication remains an invaluable lifeline in situations where immediate and reliable contact is essential.

Moreover, practical radio communication is not limited to mere voice transmission. It encompasses an array of data transmission methods, including the transmission of digital signals, allowing for the exchange of complex information, from text messages and images to video streams. This multifaceted capability has made radio communication a driving force in sectors like public safety, transportation, and military operations, where rapid data transfer is vital.

In the following exploration of practical radio communication, we will delve into its diverse applications, ranging from the intricacies of radio technology to the critical role it plays in modern society. We will also examine the challenges and innovations that have marked its evolution, and the ethical considerations surrounding its use. From ham radio enthusiasts connecting across continents to the global networks that underpin our digital lives, radio communication bridges distances and cultures, proving that even in our digital age, the airwaves continue to be a conduit of human connection, innovation, and progress.

ADVANCED RADIO ETIQUETTE AND PROTOCOLS

Practical radio communication involves the use of two-way radios for transmitting and receiving messages. Radio communication is widely used in various fields such as public safety, aviation, maritime, military, and amateur radio operation. Advanced radio etiquette and protocols ensure that communication is effective, efficient, and within the legal boundaries set by regulatory bodies. Below are some key elements:

Advanced Radio Etiquette:

1. **Identification**: Always identify yourself and the intended recipient at the beginning of the communication.

   For ham radio operators, use your assigned call sign for identification.

   In professional environments, you might use your role, station, or unit name.

2. **Speak Clearly and Slowly**: Enunciate your words clearly, avoid slang, and speak at a moderate pace to ensure understanding, especially in situations where signals may be weak or distorted.

3. **Use of Pro-Words**: Employ words like "Affirmative" for yes, "Negative" for no, "Over" to end transmission, and "Out" to end communication to reduce confusion.

   Learn and use the phonetic alphabet (Alpha, Bravo, Charlie, etc.) to spell out words or call signs.

4. **Listening First**: Always listen before transmitting to ensure the channel is clear and to avoid interrupting ongoing communication.

5. **Maintain Brief and Concise Communication**: Limit transmissions to necessary information only, especially in emergency situations where channels need to be available for critical communication.

Advanced Protocols:

1. **Emergency Protocols:**

Understand and follow procedures for emergency communication, knowing the priority of transmissions: distress, urgency, safety, and routine.

In amateur radio, emergency traffic always has priority.

2. **Repeaters and Tone Access:**

Familiarize yourself with using repeaters and CTCSS/DCS tones if necessary to access certain channels or networks.

3. **Handling Interference:**

Change your frequency or use techniques to minimize interference when experienced. Report harmful interference to the appropriate authorities.

4. **Digital Modes:**

Understand different protocols and standards like DMR, D-Star, or System Fusion when using digital modes. Each digital mode might have specific etiquette and operational procedures.

5. **Legal Compliance:**

Know and comply with the laws and regulations governing radio communication in your jurisdiction, including licensing requirements, permissible frequencies, and power limits.

Practicing Advanced Etiquette and Protocols:

Join a local radio club or group where you can practice and learn from experienced operators.

Participate in nets, which are scheduled radio meetings, to practice communication protocols and etiquette.

Engage in emergency drills or "simulated emergency tests" organized by many radio clubs to prepare for real emergencies.

REAL-WORLD COMMUNICATION SCENARIOS:

1. **Emergency Services (Police, Fire, and Medical):**
   - **Scenario**: In a multi-vehicle accident on a busy highway, first responders from different agencies need to coordinate.
   - **Communication Needs**: Quick, clear, and secure communication among diverse teams to share information about casualties, assistance needed, traffic control, and more.
2. **Aviation:**
   - **Scenario**: Air traffic controllers need to guide multiple aircraft on the ground and in the air.

- **Communication Needs**: Precise and immediate communication to convey instructions and receive acknowledgments from pilots to avoid collisions and ensure safe takeoffs and landings.

3. **Maritime Operations:**

- **Scenario**: A ship at sea needs to communicate with a coastal station or other vessels for navigation and safety purposes.

- **Communication Needs**: Long-range communication capabilities utilizing marine VHF radios and following established maritime communication protocols.

4. **Event Management:**

- **Scenario**: During a large public event like a concert or festival, organizers, security, and staff must coordinate effectively.

- **Communication Needs**: Portable radios for team communication about crowd control, logistics, emergencies, and overall event coordination.

5. **Outdoor Recreation and Expeditions:**

- **Scenario**: A group of hikers embarks on a multi-day expedition in a remote area with limited cellular coverage.

- **Communication Needs**: Radios for reliable group communication and coordination, especially in emergencies.

6. **Amateur Radio Operators (Ham Radio):**

   - **Scenario**: During a natural disaster, amateur radio operators assist with emergency communication efforts.

   - **Communication Needs**: Ham radios for a resilient, decentralized communication network when traditional networks are down or overloaded.

7. **Military Operations:**

   - **Scenario**: Military units on a mission require secure and reliable communication to coordinate maneuvers and relay critical information.

   - **Communication Needs**: Military radios with encryption and other advanced features for secure communication in various environments

Practical Communication Tips for These Scenarios:

- **Clear and Concise**: Keep messages short and to the point to ensure understanding and maintain channel availability for others.

- **Use of Codes and Phonetics**: Employ standardized codes and the phonetic alphabet to minimize misunderstandings.

- **Listen Before Speaking**: Ensure the channel is clear before beginning transmission to avoid interrupting ongoing communication.

- **Understand Priorities**: Recognize the importance of emergency traffic and give precedence when necessary.
- **Follow Legal and Ethical Standards**: Adhere to licensing regulations, respect privacy, and avoid transmitting false or misleading information.

DECODING RADIO JARGON: A SIMPLIFIED GUIDE

Understanding this radio lingo is crucial for effective communication. Below is a simplified guide to decoding some common radio jargon:

Basic Radio Terminology:

• **Affirmative**: Yes.

• **Negative**: No.

• **Over**: I have finished talking and am waiting for your reply.

• **Out**: I have finished talking; no reply is expected.

• **Stand by**: Wait, and I will get back to you.

• **Roger**: I have received and understood your message.

• **Copy**: I have received your message.

Phonetic Alphabet:

The phonetic alphabet is used to spell out words clearly over the radio, minimizing misunderstandings due to sound-alike letters.

A: Alpha	G: Golf	M: Mike	S: Sierra	Y: Yankee
B: Bravo	H: Hotel	N: November	T: Tango	Z: Zulu
C: Charlie	I: India	O: Oscar	U: Uniform	
D: Delta	J: Juliett	P: Papa	V: Victor	
E: Echo	K: Kilo	Q: Quebec	W: Whiskey	
F: Foxtrot	L: Lima	R: Romeo	X: X-ray	

Numbers:

Numbers are usually pronounced digit by digit to avoid confusion. For example, "one-five" for 15, "zero-three" for 03.

Emergency Terms:

- **Mayday**: A distress call indicating a life-threatening emergency.
- **Pan-Pan**: Urgency call signaling a pressing issue, not immediately life-threatening.
- **Securité**: Safety call used for important safety-related information.

Common Radio Jargon:

- **Break**: Indicates the separation between the address (or prefix) and the text of the message.
- **Say Again**: Please repeat your last message.
- **Wilco**: Will comply (used after receiving instructions).
- **Squelch**: A feature used to mute static noise when no communication is occurring.
- **CTCSS (Continuous Tone-Coded Squelch System)**: A tone allowing users to access specific radio channels and minimize interference.
- **Repeater**: A device that receives a signal and retransmits it to extend the communication range.
- **Simplex**: Radio communication where transmitters and receivers operate on the same frequency.
- **Duplex**: Using two different frequencies for transmitting and receiving.

Navigating the World of Radio Jargon: A Comprehensive Guide

Radio communication, with its unique lingo and jargon, can appear to be a cryptic language to newcomers. Whether you're an aspiring ham radio enthusiast, an emergency responder, or simply interested in understanding the radio chatter you hear, grasping the terminology is essential. Here, we present an extended set of tips to help you demystify the world of radio jargon and become fluent in this specialized language.

1. **Practice Regularly**: Like learning any language, practice makes perfect. To become proficient in radio jargon, it's crucial to consistently engage with and listen to radio communication. Regular interaction with the medium will expose you to the various terms, acronyms, and phrases used in specific contexts. Consider using a handheld transceiver or scanner to tune into relevant frequencies, like those used by local emergency services, to hone your listening skills.

2. **Ask for Clarification**: Don't allow personal doubt over a certain phrase to impede your capacity to speak proficiently. In situations of uncertainty, it is advisable to proactively seek clarification without hesitation. Experienced radio operators often exhibit a strong willingness to assist and support others who are new to the field, fostering a sense of cohesion and shared understanding among all participants. In crucial circumstances, it is essential to prioritize safety and efficiency by ensuring effective and unambiguous communication.

3. **Refer to Manuals and Resources**: Many radios come with instruction manuals that include glossaries or guides explaining the terminology used in radio communication. These resources can be invaluable references, providing you with a quick and comprehensive overview of the jargon specific to your radio equipment. Online resources and books dedicated to radio communication can also serve as informative guides.

4. **Join Radio Communication Groups**: Engaging with seasoned radio users can be an enriching experience. Whether you join online forums, social media groups, or local clubs and associations dedicated to radio communication, you'll have the opportunity to learn and practice radio jargon in a supportive community. These groups often host events, share knowledge, and provide hands-on training, making it an excellent way to expand your understanding of radio terminology.

5. **Participate in Simulated Exercises**: To immerse yourself fully in the world of radio communication, consider participating in simulated exercises or drills. Many emergency service agencies conduct regular training sessions where radio communication is an integral part of the exercise. This real-world practice will expose you to the actual terminology and protocols used in the field, allowing you to gain practical experience in a controlled environment.

6. **Develop a Personal Glossary**: As you encounter new terms and phrases, keep a log or journal of them. This personal glossary will serve as a handy reference that you can customize to your needs. Include definitions, context, and examples for each term to reinforce your understanding.

By adhering to these comprehensive guidelines, you will make significant progress in acquiring expertise in the intricacies of radio terminology. As one delves further into the realm of radio communication, it becomes evident that this particular lexicon amplifies one's capacity to establish connections with others, whether for recreational, emergency response, or any other reason contingent upon lucid and efficient communication.

SPEAK THE LANGUAGE

In the realm of radio communication, accurate language use is pivotal for facilitating clear and efficient exchanges. Below is a comprehensive reference designed to aid users in effectively and confidently navigating the sphere of radio communications.

Understand the Basics:

1. **Call Signs:**
   - Unique identifiers designated to each radio operator or station.
   - Amateur radio operators receive call signs from their national telecommunications authority.
   - For example, "Kilo India Six November Alpha Zulu".

2. **Phonetic Alphabet:**
   - A standardized set of words utilized to represent letters distinctly.
   - For instance, the word "RADIO" would be conveyed as "Romeo Alpha Delta India Oscar".

3. **Numerical Pronunciation:**
   - Numbers are articulated digit by digit.
   - "357" would be pronounced as "Three Five Seven".

Use Common Terms:

- "Affirmative" or "Roger": Yes or understood.
- "Negative": No.
- "Over": I have concluded speaking, and I await your response.
- "Out": Communication is concluded; no response is needed.
- "Standby": Please wait.
- "Copy": Acknowledgment of message receipt.

Emergency Codes:

- "Mayday" (Distress): Employed for life-threatening emergencies.
- "Pan-Pan" (Urgency): For urgent, non-life-threatening situations.
- "Securité" (Safety): To disseminate safety information.

Procedural Words (Pro-words):

- "Say Again": Repeat your last message.
- "Correction": An error in the transmission, the correct word is...

- "I Spell": I will spell it phonetically.

Practical Tips for Effective Communication:

1. **Clarity**: Effective communication hinges on clarity. Whether via a two-way radio or in personal interactions, precise, articulate speech is key. Enunciate each word, convey ideas effectively, and modulate your voice to impart information accurately. Avoid incoherent speech, rapid delivery, or specialized terminology that may confuse others. Precise enunciation improves communication intelligibility and reduces misunderstanding potential.

2. **Brevity**: Conciseness is vital. Verbose communication can clog channels, cause delays, and annoy others awaiting crucial information transmission. By articulating your main idea promptly and concisely, you ensure message comprehension without unnecessary distractions.

3. **Pause Before Speaking**: With Push-to-Talk (PTT) devices, observe a brief silence before initiating communication. This practice ensures your message's introductory part remains intact and informs others of your impending transmission, minimizing message interference or overlap. This fosters structured and respectful knowledge dissemination.

4. **Listening**: Communication is reciprocal, with listening as vital as speaking. Actively listen to ongoing channel discussions before transmitting. This not only provides contextual information and maintains event awareness but also prevents unintended disruptions. Interrupting ongoing discussions can hinder information exchange and cause misunderstandings. Exhibiting patience while awaiting an opportune speaking moment demonstrates respect for others and fosters a harmonious communication environment.

These practical tips, while illustrated within the realm of radio communication, are transferable to any interaction form, be it in professional settings or everyday conversations. By emphasizing clarity, brevity, thoughtful pauses, and active listening, you significantly enhance your message delivery effectiveness, contributing to smoother and more respectful information exchange in any communication scenario.

Practice Proper Etiquette:

- Wait for your turn to speak.
- Do not interrupt ongoing conversations unless there's an emergency.
- Avoid slang or unnecessary words; adhere to official radio terminology.
- Use plain language whenever possible, reserving codes and jargon for necessary situations.

Engage in Practice:

- Participate in nets or radio check-ins to practice communication.

- Join amateur radio clubs to learn from experienced operators.
- Engage in simulated emergency tests or drills.

Continuous Learning:

- Stay updated on new terms, technologies, and protocols.
- Practice regularly to become comfortable and fluent in radio communication language.

CHANNELS

When using radios for communication, understanding how channels work is crucial. Channels in radio communication refer to specific frequencies or bands of frequencies allocated for transmitting and receiving messages.

1. **Understanding Channels:**

Radio communication fundamentally relies on the allocation and usage of designated channels to facilitate the transfer of information. Channels serve as vital conduits for the transmission of data, each with distinct attributes catering to specific objectives.

- **Frequency**: Frequency is a pivotal concept in radio communication, referring to the specific rate at which electromagnetic waves oscillate for the purpose of transmitting and receiving communications. Different communication objectives necessitate the allocation of varying frequency bands. The selection of a particular frequency is often influenced by factors such as communication range, signal propagation characteristics, and regulatory limitations.
- **Channel**: In the realm of radio communication, a channel is a designated frequency or pair of frequencies assigned for communication purposes. Channels act like dedicated lanes on a highway, ensuring that multiple communication streams can coexist without interference. Typically, these channels are managed and regulated by governmental agencies to prevent overcrowding and maintain order in the radio spectrum.
- **Simplex Channel**: Simplex channels employ a single frequency for both transmitting and receiving, akin to a one-way street where traffic flows in only one direction at a time. Simplex channels are commonly used in scenarios where communication is unidirectional or when only one party needs to transmit at a given moment. Examples include walkie-talkies used for recreational purposes or in industrial settings.
- **Duplex Channel:** Conversely, duplex channels utilize different frequencies for transmitting and receiving, allowing for simultaneous communication much like a two-way street where vehicles can travel in both directions at the same time. Duplex channels are prevalent in more complex communication

41

systems, such as mobile phones and two-way radios used by emergency services. By separating transmit and receive frequencies, users can engage in full-duplex conversations, enabling real-time, two-way communication.

A thorough understanding of the various types of channels is vital for the development and execution of efficient radio communication systems. The decision between using simplex or duplex channels, as well as identifying suitable frequencies, hinges on the specific communication requirements and the operational context of the system. Grasping the nuances of channels and their characteristics is crucial to ensure reliable and interference-free communication, a critical aspect in several domains including public safety, transportation, amateur radio, and satellite communication.

2. **Types of Channels::**

- **VHF (Very High Frequency):**

  - Ranges from 30 MHz to 300 MHz.

  - Ideal for short-distance communication with less interference.

  - Commonly used in aviation, marine, and terrestrial applications.

- **UHF (Ultra High Frequency):**

- Ranges from 300 MHz to 3 GHz.

- Offers shorter wavelengths, making it suitable for urban areas with many obstructions.

- **HF (High Frequency):**

- Ranges from 3 MHz to 30 MHz.

- Suitable for long-distance communication.

- **FM and AM Channels:**

  - Frequency Modulation (FM) and Amplitude Modulation (AM) refer to the ways signals are encoded.

  - FM often provides clearer sound but has a shorter range, while AM can travel longer distances.

3. **Channel Etiquette and Protocols:**

Effective radio communication relies not only on the technology at hand but also on the adherence to a set of channel etiquette and protocols that ensure seamless and courteous interactions between users. These guidelines serve as a vital framework, guaranteeing that radio frequencies remain efficient, organized, and accessible to all who require them. Whether you're a novice or a seasoned radio operator, understanding and respecting these principles is paramount for harmonious and productive radio communication.

- **Listen Before Transmitting**: One of the cardinal rules of radio communication is the practice of listening before transmitting. Before initiating any communication, it is essential to monitor the channel for ongoing conversations. This serves a dual purpose – it helps you determine if the channel is available for your use and ensures you don't inadvertently interrupt ongoing discussions. The importance of this cannot be overstated, as barging into an active conversation can not only be disruptive but also hinder the exchange of critical information.

- **Avoid Channel Congestion**: Radio channels are shared resources, and as such, it is incumbent upon all users to keep their transmissions concise and to the point. Verbose or unnecessary chatter can lead to channel congestion, making it difficult for others to get their messages across. To maintain effective communication, it is advisable to communicate your message as succinctly as possible. This not only respects the time and resources of other users but also ensures that the channel remains available for everyone when needed.

- **Use Appropriate Channels**: Different radio channels are designated for specific purposes, such as emergency, marine, aviation, and more. To avoid confusion and maintain order, it is imperative to use the correct channel for your communication needs. Misusing channels not only violates established protocols but can also have serious consequences, particularly in critical situations. Being aware of and adhering to these designations is essential for efficient communication and, in some cases, for ensuring safety.

These channel etiquette and protocols serve as the unwritten code that governs the proper use of radio communication. They are not arbitrary rules but are rooted in a deep understanding of the technology's limitations and the necessity for clear, organized, and interference-free transmissions. By following these guidelines, radio users can ensure that their messages are transmitted and received accurately, without unnecessary disruptions, contributing to the overall effectiveness and reliability of radio communication networks.

Channel etiquette and protocols are the cornerstone of effective radio communication. They ensure that the airwaves remain open and accessible to all who rely on them, promoting clarity, efficiency, and safety in this vital mode of communication. Whether you're a hobbyist or a professional, mastering these guidelines is a fundamental aspect of responsible and effective radio operation.

4. Programming and Selecting Channels:

Efficient and accurate channel selection and programming are pivotal aspects of radio communication systems. Whether you're using radios for personal use, public safety, or professional applications, the ability to manage your channels effectively ensures smooth and reliable communication. Here, we explore several methods for programming and selecting channels, catering to a wide range of user preferences and requirements.

- **Manual Programming**: One of the most straightforward methods for channel selection is manual programming. This approach allows users to manually tune or select specific frequencies or channels on their radios. Manual control is especially useful when users need to fine-tune their communication channels, switch between different frequencies, or adapt to rapidly changing conditions in the field. While it may require a bit more user effort, manual programming offers a high degree of flexibility.

- **Software Programming**: Software-based channel programming has gained popularity for its efficiency and ease of use. Applications like CHIRP provide a user-friendly interface for programming a wide range of radios. Users can connect their radios to a computer and utilize the software to manage and configure multiple channels simultaneously. This method is particularly advantageous for organizations or individuals who need to maintain and update a large fleet of radios, as it streamlines the process and reduces the likelihood of programming errors.

- **Presets:** Many modern radios come equipped with preset channels, making them user-friendly and accessible, even for beginners. These presets include commonly used frequencies for various purposes, such as emergency services, weather updates, or specific industries. The advantage of preset channels is their convenience, as users can quickly access vital information or switch to relevant channels without the need for manual programming. Presets enhance the user experience and ensure that important channels are readily available.

In addition to these primary methods, some advanced radio communication systems offer features like dynamic channel selection and group-based programming, which further streamline the channel management process. These advanced capabilities are particularly valuable for applications requiring real-time adaptation to changing conditions or for coordinating communication within large teams.

Effective channel programming and selection are essential for optimizing radio communication systems' performance and ensuring seamless connectivity. The choice of method—whether manual, software-based, or relying on presets—largely depends on the specific needs, technical expertise, and operational requirements of the user or organization. Regardless of the chosen approach, the ultimate goal is clear: to make radio communication a reliable and efficient means of connecting people and information, regardless of the setting or context.

5. Legal Considerations:

In the intricate web of radio communication, several legal considerations require the acute awareness of users and operators to ensure compliance with established laws and regulations. These legal aspects are paramount as they not only safeguard the orderly and efficient use of the radio spectrum but also protect the rights and interests of individuals, organizations, and the general public. Let's delve deeper into these crucial legal considerations:

- **Licensing Requirements**: Licensing is a crucial legal prerequisite in the realm of radio transmission. In several nations, certain frequencies are allocated for particular purposes, necessitating individuals or organizations intending to broadcast on these frequencies to obtain a license. Typically, these licenses are provided by the national telecommunications authority or regulatory body, aiming to ensure efficient management of the radio spectrum and thereby mitigating interference and disputes among users. The issue of licensing holds significant relevance within the sphere of radio amateurs, commercial radio operators, and entities that depend on radio frequencies for their day-to-day activities.

- **Restrictions on Frequency Usage**: Radio frequencies are a finite and valuable resource, and as such, they are carefully allocated and regulated to ensure fair and equitable use. Restrictions are imposed on the use of certain channels to prevent overcrowding and interference. For instance, specific frequencies may be reserved for emergency services, aviation, or maritime communication. These restrictions help maintain the integrity of radio communication, ensuring that critical transmissions are not hindered by unnecessary interference.

- **Compliance with Regulations**: Adherence to laws is a fundamental aspect of responsible radio transmission. The rules in question are set and implemented by the telecommunications body that has control over a certain area. The outlined provisions include the authorized applications of radio frequencies, prescribed power thresholds for transmitters, technical specifications for equipment, and further operating prerequisites. Adhering to these laws is not just a legal requirement but also a pragmatic need to guarantee the reliability, efficiency, and security of radio transmission.

Legal considerations in radio communication encompass licensing, usage restrictions, and compliance with regulations. By obtaining the necessary licenses, respecting frequency restrictions, and adhering to regulatory guidelines, radio users can contribute to the orderly and effective use of the radio spectrum. These legal requirements also serve to protect against interference, promote fair access to radio frequencies, and facilitate the coexistence of diverse radio communication services, all while safeguarding the interests of both users and the broader public. Understanding and upholding these legal aspects is a fundamental responsibility for anyone engaged in radio communication, from amateur radio enthusiasts to professional operators and organizations relying on this critical medium for their daily activities.

6. Safety and Privacy:

In the domain of radio communication, the preservation of safety and privacy holds utmost significance. Knowledge of the protocols and features that enhance safety and safeguard privacy can significantly impact the communication experience for those engaged in amateur radio operation, professionals working in the field, or leisure users.

- **Monitoring for Safety**: A fundamental aspect of radio communication safety is the practice of regularly monitoring channels for emergency transmissions and important announcements. This practice is especially crucial in scenarios where an instantaneous response is vital, such as in search and rescue operations, public safety organizations, and marine and aviation communications. By consistently monitoring radio frequencies, users can swiftly respond to distress calls, ensuring the safety of those in need. It also allows individuals to stay informed about rapidly changing situations, such as weather alerts or traffic updates, contributing to overall safety.

- **Privacy Codes (CTCSS/DCS)**: Radio channels are often shared by numerous users, and this sharing can lead to interference and crosstalk. Privacy codes, such as the Continuous Tone Coded Squelch System (CTCSS) and Digital-Coded Squelch (DCS), are invaluable tools for minimizing interference and enhancing privacy in radio communication. It's essential to understand that privacy codes do not make channels private in the strictest sense; instead, they add a layer of selective muting to radios, preventing them from responding to signals that lack the corresponding code.

CTCSS and DCS work by transmitting a sub-audible tone or digital code along with the audio signal. Radios equipped with these codes will remain silent unless they receive a signal with the correct code. This selective muting effectively reduces unwanted chatter and interference on shared channels, allowing users to communicate more clearly and with reduced disruptions.

While privacy codes may not provide absolute privacy, they significantly improve the quality of communication in busy radio environments. They are widely used in settings such as businesses, amateur radio clubs, and recreational outdoor activities, where multiple users share the same frequency. By implementing privacy codes, radio users can streamline their conversations, reduce confusion, and maintain a level of privacy that is often adequate for their needs.

Safety and privacy are integral aspects of radio communication, and adopting best practices like regular channel monitoring and implementing privacy codes can greatly enhance the quality and effectiveness of communication. These measures not only contribute to smoother and more secure interactions but also ensure that radio communication remains a reliable and valuable tool in various professional and recreational applications.

7. Emergency Channels:

When it comes to radio communication, few aspects are as crucial as the availability of dedicated channels for emergencies. These channels serve as lifelines in times of crisis, enabling swift and effective communication for those in need. Emergency channels come in various forms, each designed to meet the unique demands of specific scenarios, and they play an instrumental role in maintaining safety and connectivity in critical situations.

- **Dedicated Channels**: One of the most fundamental aspects of emergency radio communication is the existence of dedicated channels. These channels are specifically reserved for urgent situations, and their importance cannot be overstated. For instance, in the maritime world, channel 16 on the marine VHF radio is universally recognized as the international hailing and distress frequency. It serves as a beacon of hope for vessels in distress, allowing them to call for assistance and coordinate rescue efforts. These channels are monitored by authorities and first responders, ensuring that distress signals do not go unanswered.

- **Weather Channels:** Weather channels are crucial to emergency radio communication. Broadcasts from the National Weather Service are available on many radios. These channels provide live weather updates, severe weather alerts, and predictions. They offer vital information to communities, emergency services, and disaster response teams in hurricane-prone, tornado-prone, and wildfire-prone areas. Monitoring quickly changing weather conditions enables people to make safe choices for themselves and their communities.

- **Public Safety and First Responder Channels**: Beyond dedicated and weather channels, there are specialized frequencies allocated for public safety and first responder communication. Police, fire departments, and medical services rely on these channels to coordinate their efforts during emergencies. These radio systems are meticulously designed to ensure reliable, secure, and interoperable communication, allowing different agencies to work together seamlessly in complex and high-pressure situations.

- **Amateur Radio Emergency Service (ARES):** The world of amateur radio enthusiasts also contributes significantly to emergency communication. ARES groups are comprised of skilled radio operators who volunteer their time and expertise during crises. They often operate on designated amateur radio frequencies, providing backup communication for emergency services when traditional infrastructure fails.

In essence, emergency channels in radio communication are the backbone of disaster management and public safety. They serve as a lifeline, connecting those in distress with the assistance they need and facilitating the coordination of critical response efforts. Whether it's a distress call at sea, a tornado warning in a community, or the seamless collaboration of first responders, these channels exemplify the power of radio communication in safeguarding lives and property during the most challenging moments.

8. **Practical Usage Tips:**

- **Familiarize Yourself with Channel Selection**: Understanding the channels available on your radio communication device is the first step toward efficient and effective communication. Each channel is designated for specific purposes, such as public safety, marine, aviation, or amateur radio bands. It is crucial to know which channel is appropriate for your intended use. Different channels are allocated for various services to prevent interference and ensure that each communication remains clear and secure. Take the time to learn about the regulations governing the use of these channels in your region, as they may differ.

- **Label and Program Your Channels**: To streamline your radio communication experience, consider programming and labeling your channels. This step is particularly crucial in situations where time is of the essence, such as emergencies. By assigning clear labels to your channels, you can quickly access the right channel without fumbling through numerous options. For example, if you frequently communicate with a particular group or department, label a channel accordingly, so you can switch to it instantly when needed.

- **Regular Practice and Proficiency**: Like any skill, radio communication proficiency comes with practice. Regularly engaging in exercises or simulations can significantly enhance your ability to use different channels effectively. Practice not only includes learning how to switch between channels but also understanding the protocols and language conventions associated with each channel. For instance, emergency channels may have specific codes or procedures that differ from standard communication channels. Familiarizing yourself with these nuances can be a lifesaver in critical situations.

- **Learn Emergency Procedures**: In the realm of radio communication, emergencies can arise, and being prepared is paramount. Take the time to learn the emergency procedures relevant to your device and the channels you use. This includes understanding distress signals, emergency call codes, and how to respond when others are in need of assistance. Quick and precise action during emergencies can be the difference between a successful resolution and a potential disaster.

- **Maintain Your Equipment**: Radio communication devices are reliable tools, but they require proper maintenance to ensure they function when you need them most. Regularly inspect your equipment for wear and tear, replace batteries as needed, and keep your device clean and free from debris that might affect performance. Having backup batteries and other essential accessories on hand is a wise practice, particularly for extended missions or emergencies.

In conclusion, effective radio communication is not just about having the right equipment; it's about knowing how to use it efficiently and responsibly. By familiarizing yourself with channel selection, labeling channels, practicing regularly, learning emergency procedures, and maintaining your equipment, you can maximize the utility and reliability of your radio communication tools, whether in your daily activities or during critical situations where clear and swift communication is of paramount importance.

BASIC RADIO ETIQUETTE

In any form of radio communication, adhering to a set of etiquette standards is crucial. Following these norms ensures that communication is efficient, respectful, and within legal boundaries.

1. **Clear Identification:**
   - Always start a transmission by identifying yourself and the party you intend to communicate with.
   - Use assigned call signs for identification.
   - Example: "Control, this is Alpha One. Over."

2. **Brief and Clear Messages:**
   - Keep your messages short, straightforward, and to the point.
   - Avoid using filler words and slang.
   - Speak clearly and slowly to avoid misunderstandings.

3. **Proper Use of Pro-Words:**
   - Use procedural words (pro-words) to streamline communication.
   - "Over" signals the end of your transmission and invites a response.
   - "Out" indicates the end of your transmission, with no response needed.
   - "Copy" or "Roger" to acknowledge receipt and understanding of a message.

4. **Phonetic Alphabet:**
   - When spelling out words, use the NATO phonetic alphabet to avoid confusion.
   - Example: "Romeo Alpha Delta India Oscar" for RADIO.

5. **Listening Before Transmitting:**
   - Always listen to ensure the channel is clear before starting your transmission.
   - Avoid interrupting ongoing conversations unless there is an emergency.

6. **Acknowledging Messages:**
   - Respond to transmissions directed at you promptly.

   - Acknowledge receipt and understanding of messages.

7. **Emergency Protocols:**
   - Familiarize yourself with and respect emergency protocols.

   - Give priority to emergency communications.

   - Example: "Mayday" for distress calls.

8. **Avoiding Unnecessary Transmissions:**
   - Minimize transmission time to keep channels clear for others, especially during emergencies.

   - Refrain from transmitting non-essential or casual messages on official or emergency channels.

9. **Respecting Privacy:**
   - Avoid sharing sensitive personal information over the radio, as channels are not private.

   - Respect the privacy of others by not eavesdropping on conversations that don't involve you.

10. **Legal Compliance:**
    - Understand and adhere to regulations governing radio communication in your jurisdiction.

    - Operate within assigned frequencies and adhere to licensing requirements.

11. **Professionalism and Courtesy:**
    - Maintain a professional and respectful tone during transmissions.

    - Refrain from using offensive language or engaging in arguments over the radio.

12. **Battery Conservation:**
    - Turn off your radio or switch to a low power setting when not in use to conserve battery.

SCAN THE QR CODE THERE TO ACCESS THREE UNIQUE BONUSES TAILORED EXCLUSIVELY FOR YOU!

CHAPTER 4: ADVANCED RADIO COMMUNICATION

Advanced Radio Communication refers to the sophisticated, cutting-edge techniques and technologies that facilitate wireless transmission and reception of signals for communication purposes. This domain encompasses various forms of radio waves, protocols, and technologies designed to enhance and improve traditional radio communication methods.

IMPORTANCE OF ADVANCED RADIO COMMUNICATION:

1. **Emergency Response & Public Safety**: In crises such as natural disasters, terrorist attacks, or other emergencies, advanced radio communication systems are crucial for first responders and public safety officials. They provide reliable, real-time communication, often with enhanced features like encryption and GPS location.

2. **Defense & Military Applications**: Military operations often rely heavily on secure, robust, and resilient communication systems. Advanced radio communication is imperative for command and control, intelligence, surveillance, reconnaissance, and seamless coordination between various units and allied forces.

3. **Mobility & Connectivity:** Advanced radio communication systems support the connectivity needs of a mobile, dynamic society. They enable mobile phones, Wi-Fi networks, satellite communications, and other wireless technologies, allowing people to stay connected irrespective of their location.

4. **Communication in Remote Areas**: In areas where wired communication infrastructure is not feasible due to geographical challenges or economic reasons, advanced radio communication can provide a lifeline, enabling connectivity and access to essential services.

5. **Scientific Research & Exploration**: Scientists and researchers use advanced radio communication for purposes like space exploration, deep-sea research, and environmental monitoring. These systems can transmit data over long distances, often in harsh or inaccessible environments.

6. **Economic Development & Infrastructure**: Many sectors, including transportation, logistics, agriculture, and healthcare, depend on advanced radio communication. For instance, smart grids rely on wireless communication for monitoring and control. Similarly, automated vehicles, drones, and IoT devices utilize these systems for connectivity and operation.

7. **Innovation & Technological Advancements:** The field of radio communication is continually evolving, with new technologies like 5G, Low Earth Orbit (LEO) satellites, and software-defined radios

emerging. These innovations offer improved performance, efficiency, and capabilities, opening up new possibilities and applications.

Advanced Radio Communication Technologies:

- **Software-Defined Radio (SDR):** SDR allows for flexibility and versatility in handling different communication protocols and frequencies. This technology can be reprogrammed and reconfigured through software, making it adaptable to various applications and standards.
- **Digital Radio:** Unlike analog, digital radio offers clearer audio quality, better signal integrity, and the ability to transmit data alongside audio. This technology also supports advanced features like digital encryption and error correction.
- **Satellite Communication**: Satellites facilitate global communication, providing coverage in remote and inaccessible areas. They support various services, including broadcasting, internet connectivity, navigation, and scientific research.
- **5G Technology**: The fifth generation of cellular network technology, 5G, offers faster speeds, lower latency, and the ability to connect a significantly higher number of devices concurrently. This technology is crucial for emerging applications like autonomous vehicles, smart cities, and augmented reality.

Repeater

A radio repeater is a crucial component in advanced radio communication systems, extending the range and effectiveness of wireless communication by receiving, amplifying, and retransmitting signals. Repeaters are predominantly used in two-way radio systems and networks, including public safety, commercial, and amateur radio services. Below is a detailed discussion about the function and importance of radio repeaters in advanced radio communication systems:

Functions of Radio Repeaters:

1. **Signal Amplification**: Repeaters receive weak or low-level radio signals and amplify them before retransmitting. This is vital for maintaining the clarity and strength of communication over long distances.
2. **Range Extension**: They significantly extend the communication range of radio devices. Without repeaters, radio signals might be limited due to geographical obstructions, signal attenuation, or the inherent limitations of the transmitting devices.
3. **Overcoming Obstructions**: Radio signals can be obstructed or reflected by buildings, mountains, or other structures. Repeaters, when strategically located, can help in overcoming these barriers to provide seamless communication.

Components of Radio Repeaters:

1. **Receiver**: The receiver captures incoming radio signals. It is designed to be sensitive enough to pick up weak signals while maintaining a high degree of selectivity to ignore out-of-band interference.

2. **Transmitter**: The transmitter is responsible for sending out the amplified signals. It should be powerful enough to cover the desired service area without causing interference to other communication systems.

3. **Duplexer**: A duplexer allows the repeater to transmit and receive signals simultaneously on a single antenna while preventing the transmitter from overpowering the receiver. This component is crucial for the repeater's full-duplex operation.

4. **Controller**: The controller is the brain of the repeater, managing its functions and features. It may provide additional services like identifying the repeater station, managing talk time, and handling emergency signals.

Types of Radio Repeaters:

- **Analog Repeaters**: These are traditional repeaters that work with analog signals. While they are simpler and more robust, they might lack the advanced features and functionalities provided by their digital counterparts.

- **Digital Repeaters:** Digital repeaters work with digitally encoded signals, offering better signal integrity, security, and additional features like text messaging, GPS, and emergency alerts.

Importance of Radio Repeaters:

1. **Emergency Communication**: In disaster recovery or emergency response situations, communication is often disrupted. Repeaters play a critical role in maintaining and establishing communication networks under these circumstances.

2. **Public Safety:** Police, fire, and other public safety agencies rely on two-way radio systems with repeaters for reliable communication, essential for coordinating efforts and responding to incidents effectively.

3. **Commercial Applications**: In large facilities, construction sites, or transportation networks, repeaters ensure that communication is possible over large areas or between vehicles.

4. **Amateur Radio**: Ham radio operators often use repeaters to communicate over long distances, connect to the Internet, or link with other amateur radio networks.

DATA HANDLING

Data handling in the context of advanced radio communication refers to the processes involved in transmitting, receiving, and managing data efficiently over radio frequencies (RF). With the advent of digital technology, radio communication systems have evolved beyond voice to support the transmission of various forms of data, including text, images, video, and more. Below are key elements and considerations related to data handling in advanced radio communication:

Key Elements:

1. **Modulation and Demodulation:**
   - **Modulation**: At its core, modulation is the clever method of transforming digital data into radio waves. Fundamentally, this refers to the process through which binary data, often denoted as 0s and 1s, is converted into electromagnetic radiation capable of propagating through the atmosphere. A multitude of modulation schemes are used to accomplish this objective, each with distinct attributes and practical implementations. Quadrature Amplitude Modulation (QAM), Phase Shift Keying (PSK), and Frequency Shift Keying (FSK) are often used modulation techniques.
   - **Demodulation**: The counterpart to modulation, demodulation is the process of extracting the original data from the modulated signals that are received. It is a crucial step in radio communication, as it essentially reverses the encoding process, allowing us to recover the digital information sent by the transmitter.

2. **Error Detection and Correction:**
   - To ensure data integrity, error detection and correction techniques are implemented. These mechanisms identify and correct errors that may occur during transmission, often due to interference, noise, or signal fading.

3. **Data Compression:**
   - Compression reduces the size of data to be transmitted, facilitating efficient utilization of bandwidth. Various algorithms and protocols are employed to compress data without significant loss of quality.

4. **Data Security and Encryption:**
   - Security is paramount, especially for sensitive or confidential information. Data transmitted over radio waves is often encrypted to prevent unauthorized access and ensure confidentiality and integrity.

5. **Protocols and Standards:**
   - Communication protocols and standards define the rules and formats for data transmission and reception. These guidelines ensure compatibility, interoperability, and reliable communication between different devices and systems.

Considerations:

1. **Data Rate**: The data rate, or throughput, refers to the amount of data transmitted per unit time. Advanced radio communication systems aim to maximize data rates while maintaining signal quality and reliability.

2. **Latency**: Latency is the delay between transmitting and receiving data. Low latency is crucial for applications requiring real-time communication, like video conferencing, gaming, and autonomous vehicles.

3. **Bandwidth**: Bandwidth refers to the range of frequencies available for data transmission. More bandwidth allows for higher data rates and the simultaneous transmission of multiple signals.

4. **Spectrum Management**: Efficient spectrum management ensures optimal utilization of available frequency bands, minimizing interference and promoting coexistence of multiple communication systems.

5. **Quality of Service (QoS)**: QoS mechanisms prioritize different types of data, allocating bandwidth and resources accordingly. This is essential for ensuring uninterrupted service for critical applications.

Applications:

1. **IoT Devices**: The proliferation of Internet of Things (IoT) devices is occurring at a remarkable pace, as they become more integrated into various aspects of our residences, urban environments, and economic sectors. These gadgets primarily depend on radio communication for the transmission and reception of data. Efficient data management plays a crucial role in the operation of linked devices and intelligent systems, including applications such as smart thermostats for temperature regulation in residential settings, industrial sensors for monitoring equipment performance, and agricultural sensors for improving crop development. The implementation of practical radio communication facilitates the smooth exchange of information between devices, as well as with central control systems. This capability enables the automation of processes, the analysis of data, and ultimately leads to enhanced decision-making.

2. **Telecommunications**: The development of the telecommunications industry can largely be attributed to the use of modern radio communication technologies. Cellular networks employ complex radio communication technologies to provide voice, text, and internet services to mobile devices. The cornerstone of providing fast and dependable connections to billions of consumers globally lies in the effective management of data. The progression of radio communication from 2G to 5G and beyond has facilitated the uninterrupted flow of data that sustains our digital existence. These networks have emerged as the fundamental framework of our global communication infrastructure, serving as the foundation for both individual communication and the operations of enterprises and governmental entities.

3. **Broadcasting**: Radio broadcasting has a storied history, and it has gracefully adapted to the digital age. Digital broadcasting services transmit audio, video, and multimedia content to audiences using various modulation schemes and data handling techniques. From FM and AM radio to television broadcasts and

streaming services, radio communication remains at the core of our media consumption. It ensures that we can receive information and entertainment with clarity and consistency, even as we transition from traditional airwaves to online platforms.

4. **Navigation and Tracking**: Satellite-derived navigation and tracking technologies, exemplified by GPS (Global Positioning System), GLONASS, and Galileo, have brought about a paradigm shift in global navigation methodologies. These systems are dependent on radio communication for the transmission of position data and other relevant information between satellites and devices. Practical radio communication plays a crucial role in several applications, such as facilitating cross-country road trips, enabling precise precision farming, and assuring the safety of sea navigation. It allows for the accurate pinpointing of locations and tracking of movement with exceptional precision.

5. **Emergency Services**: First responders and emergency services are often the unsung heroes in times of crisis. They depend on reliable, secure radio communication systems for coordination and operation during emergencies. Police, fire, and medical professionals use radio communication to swiftly and effectively respond to incidents, ensuring public safety. From natural disasters to medical emergencies, radio communication is the lifeline that allows these brave individuals to act decisively and save lives.

BENEFITS OF USING ADVANCED REPEATER SERVERS

In the realm of advanced radio communication, repeater servers play a significant role by extending and enhancing wireless communication capabilities. They primarily receive radio signals, amplify them, and retransmit them to cover a larger geographical area, providing a myriad of benefits to users.

Enhanced Coverage:

1. **Extended Range:**
   - Repeater servers significantly boost the range of radio signals, facilitating communication over extended distances. This feature is crucial in vast or remote areas where signal reach is inherently limited.
   - In urban environments with many obstructions, repeaters help in providing consistent coverage.

2. **Overcoming Geographical Challenges:**
   - The presence of hills, buildings, or other obstacles can hinder or block radio signals. Strategically located repeater servers can bypass these obstructions to maintain seamless communication.

3. **Indoor and Underground Coverage:**
   - In scenarios where signals need to penetrate through buildings or reach underground facilities, advanced repeater servers help in providing the necessary coverage.

Improved Communication Quality:

1. **Signal Amplification:**

In the realm of communication technology, signal strength is a critical factor influencing the clarity and reach of our messages. Weak signals, which often result from a variety of factors such as geographical barriers or obstacles, can lead to disruptions, dropped calls, and frustratingly slow data transfers. To address this challenge, modern communication systems employ servers that possess the remarkable capability of signal amplification.

These servers serve as the backbone of signal strength enhancement. They work by identifying incoming signals, detecting any weakening that may have occurred during transmission, and subsequently amplifying them. The result is a substantial boost in signal strength, ensuring that the end-users receive clear, uninterrupted, and high-quality communication. The amplification process significantly improves voice call quality, data transfer rates, and overall user experience.

2. **Noise Reduction:**

In the midst of our bustling, technologically infused lives, interference and noise can be constant companions, disrupting the pristine quality of our communications. Be it the static on a radio station, background noise during a phone call, or interference from nearby electronic devices, noise can be a persistent annoyance that hampers effective communication. To combat this, advanced signal processing techniques have been developed to address the issue of noise reduction.

These signal processing techniques implemented within communication servers are designed to distinguish between the intended signal and unwanted interference. They meticulously analyze incoming data, identifying noise and interference patterns. Once detected, they utilize algorithms to filter out these unwanted elements, leaving behind the clean, undistorted signals. This process leads to a significant reduction in noise, ensuring that the recipient experiences clear and reliable communication, free from the distractions that noise can introduce.

Efficient Network Management:

1. **Traffic Handling**: Advanced repeater servers can manage communication traffic efficiently by prioritizing signals and handling multiple channels to accommodate various users without congestion.
2. **Resource Allocation**: Dynamic resource allocation allows for optimized use of available frequencies and channels, adapting to the changing demands and loads on the network.

Enhanced Features:

1. **Digital and Analog Compatibility**: Modern repeater servers often support both analog and digital signals, providing flexibility and facilitating the transition between different technologies.

2. **Advanced Protocols**: They can handle advanced communication protocols, including error correction, data compression, and encryption, improving the overall performance and security of radio communication networks.

3. **Integration with IP Networks**: IP connectivity allows for integration with existing networks and infrastructure, expanding communication possibilities and enabling features like remote control and monitoring.

Improved Reliability and Resilience:

1. **Redundancy and Backup**: By implementing redundant systems and backups, advanced repeater servers ensure uninterrupted service even in the case of hardware failures or other issues.

2. **Emergency and Disaster Response**: During emergencies, reliable communication is vital. Repeater servers can be quickly deployed or activated to establish communication networks for first responders and rescue teams.

Cost-Efficiency:

1. **Infrastructure Savings**: One of the paramount advantages of optimizing radio communication systems lies in the substantial infrastructure savings it offers. By extending the range and improving the overall efficiency of radio networks, organizations can dramatically curtail the need for additional infrastructure investments. The expansion of radio coverage through smart network planning, more efficient spectrum utilization, and advanced technologies means that businesses and institutions can serve a broader area with the same or even fewer base stations and transmission facilities. This, in turn, leads to a remarkable reduction in capital expenditure. Fewer towers, transmitters, and receivers not only decrease the initial investment required but also lower ongoing operational costs, including energy consumption, site maintenance, and rental fees for tower space. In an era where every penny saved matters, the cost-effectiveness of optimizing radio infrastructure cannot be overstated.

2. **Low Maintenance**: The adoption of advanced repeater servers represents another significant facet of cost-efficiency in the realm of radio communication. These repeater servers, with their intelligent design and robust engineering, are tailored to operate with minimal maintenance requirements. Their reliability ensures that organizations can count on consistent and dependable performance without the need for frequent and costly upkeep. Reduced maintenance efforts translate directly into cost savings, not only in terms of money spent on repairs and replacements but also in the invaluable resource of time. With less

time dedicated to system maintenance, staff can focus on more strategic tasks, enhancing overall productivity. Furthermore, the extended lifespan and enhanced reliability of low-maintenance radio infrastructure mean a lower total cost of ownership, enabling organizations to allocate their budgets more effectively and sustain their radio communication networks with confidence.

In conclusion, the cost-efficiency associated with optimizing radio communication systems transcends the mere fiscal realm. It not only conserves financial resources but also bestows organizations with a sense of operational freedom and reliability. Reduced infrastructure costs and low maintenance requirements are just two dimensions of this multifaceted advantage. By embracing advanced technologies and smarter planning, businesses and institutions can unlock significant savings, enabling them to allocate their resources more effectively and concentrate on what truly matters - seamless, efficient, and dependable communication.

CHAPTER 5: MASTERING RADIO PROGRAMMING

Mastering radio programming refers to becoming proficient in creating, scheduling, and managing content that is broadcast over radio waves to listeners. The radio industry has undergone significant changes over the years, with digital technologies and streaming platforms emerging, but the fundamental skills required for radio programming remain relatively consistent. Here are some key components and steps to mastering radio programming:

1. **Understanding the Basics:**

To embark on a journey into the dynamic world of radio communication, it is essential to develop a solid foundation in its fundamental principles and tools. The following points provide an overview of the key aspects one should grasp when seeking to comprehend and engage with radio communication effectively.

- **Knowledge of Radio Formats**: Radio communication is a multifaceted medium that encompasses a diverse array of forms, each possessing distinct qualities and necessitating specific needs. It is necessary to undertake a comprehensive examination and comprehension of these many forms, including but not limited to discourse, musical compositions, journalistic content, and athletic events. The aforementioned formats function as the fundamental components of radio programming, each necessitating distinct material, tone, and style. Acquiring proficiency in these domains is crucial for anyone with aspirations of becoming radio presenters, producers, or content creators.

- **FCC Regulations**: The Federal Communications Commission (FCC) serves as the regulatory authority for radio communication in the United States. It enforces rules and guidelines that govern radio broadcasts, ensuring that they meet technical and content-related standards. A thorough familiarity with FCC regulations is paramount for anyone involved in radio, as non-compliance can lead to penalties and disruptions in broadcasting. Understanding issues related to broadcasting licenses, signal strength, and indecency regulations is vital to navigate the complex regulatory landscape.

- **Audio Editing Skills**: The production and delivery of polished, entertaining, and professional audio are characteristic of radio. This requires audio editing and mixing skills using software like Audacity, Adobe Audition, Pro Tools, or similar platforms. These technologies enable radio professionals to record, edit, and improve audio material by removing background noise, adding sound effects, and enhancing sound quality. These skills allow radio producers to create excellent programming and adapt to changing broadcast needs and listener expectations.

Grasping the basics of radio communication involves more than just knowing how to speak into a microphone. It requires a deep understanding of various radio formats, compliance with regulatory standards set by organizations like the FCC, and the ability to craft and refine audio content through skillful editing. These foundational elements are the building blocks upon which aspiring radio professionals can construct a successful and fulfilling career in the exciting and ever-evolving world of radio communication

2. **Content Creation:**

- Content Planning: The initial critical stage in content development is planning. This stage necessitates the construction of an editorial schedule, selection of subjects, and careful delineation of segments that will captivate your target audience. A well-planned editorial schedule ensures a steady flow of fresh and pertinent material, aiding in maintaining consistency which is crucial for cultivating a loyal audience. By understanding your audience's interests, you can craft material that entices them to return. Effective content preparation allows you to delve into multiple topics, keeping radio programs exhilarating and versatile.

- Scriptwriting: Upon crafting a content outline, proceed with scriptwriting. Radio scriptwriting demands creativity and precision. To accommodate time and broadcast style, these scripts must be concise, engaging, and well-organized. Adept scriptwriters strike a balance between informing and entertaining their audiences. Effective scriptwriting is crucial for creating valuable, emotionally resonant material that connects with listeners. A compelling script is essential for a strong radio presence.

- Voice Modulation: Voice modulation, the final step in content development, assists in communicating clearly, expressively, and emphatically. Radio broadcasters must master voice modulation to communicate successfully and retain listener interest. Voice modulation, crucial for audience engagement, allows you to convey a range of emotions, keeping your broadcast lively. It helps articulate the right emotions, emphasize key points of your material, and ensure your message resonates, whether you're sharing a heartwarming story, addressing a grave issue, or inciting laughter.

3. **Audience Engagement:**

- **Understanding the Audience**: Analyze your audience's demographics, preferences, and listening habits to tailor content accordingly.

- **Social Media Integration**: Engage listeners through social media platforms, promote content, and foster a community around your programming.

4. **Technology Mastery:**

- Broadcast Equipment: It's advisable to familiarize oneself with the hardware and software utilized in radio broadcasting, including devices such as microphones, mixing consoles, and automation systems.

- Streaming Platforms: Understand how to distribute content over the internet, encompassing podcasts and live streaming.

5. **Continuous Improvement:**

- Feedback Mechanism: Establish methods to receive and analyze feedback from listeners and stakeholders for continuous improvement.

- Market Trends: Stay abreast of industry trends, emerging technologies, and shifts in listener behavior to remain relevant and competitive.

6. **Networking:**

- Industry Connections: Forge relationships with other radio professionals, content creators, and industry experts to broaden your network and opportunities.

7. **Professional Development:**

- Engage in ongoing learning through workshops, seminars, conferences, and other professional development opportunities.

Practical Steps to Mastery:

Embarking on the path toward radio expertise is a rewarding venture that demands dedication, skill cultivation, and a genuine passion for the medium. While it might initially appear daunting, a set of pragmatic steps can offer guidance in honing one's skills and achieving success within the realm of radio broadcasting. To enhance one's proficiency as a radio broadcaster, it's crucial to thoroughly explore the fundamental stages involved, thereby providing a comprehensive perspective on the process of realizing one's full potential.

- **Start Small**: It's often said that every accomplished radio personality began their career with small yet significant steps. Volunteering at local community or college radio stations can serve as the perfect entry point. This hands-on experience allows you to familiarize yourself with the technical aspects of radio, practice using the equipment, and most importantly, get comfortable behind the microphone. Whether you're hosting a talk show, spinning tunes, or providing the community with news updates, this foundation will prove invaluable as you advance.

- **Experiment**: Radio is a versatile medium, offering a multitude of formats, styles, and content types. To discover what resonates best with both you and your audience, don't hesitate to experiment. Explore different show formats, from talk radio and music programs to podcasts and live interviews. Understanding the diverse options available will help you find your unique voice and niche in the radio landscape.

- **Mentorship**: Learning from those who have already traversed this path can significantly accelerate your journey toward radio mastery. Seek out seasoned radio programmers and professionals who can offer guidance, share their expertise, and provide invaluable insights. A mentor can assist you in navigating the industry, honing your skills, and averting common pitfalls. Their wisdom can serve as a treasure trove of knowledge and inspiration.

- **Consistency**: Success in radio hinges on consistency. Building an engaged audience requires regular content creation and distribution. Maintain a broadcast schedule that suits you. Whether it's a weekly program, daily podcast, or anything in between, your audience should anticipate your material at certain times. Engaging with listeners through social media, emails, or on-air call-ins fosters community and loyalty.

In your radio journey, remember that mastery is an ongoing process. While these practical steps lay a strong foundation, your growth as a radio broadcaster will stem from a blend of hands-on experience, exploration, learning from mentors, and maintaining a steadfast commitment to your craft. Over time, your unique style and voice will emerge, attracting an audience that values your expertise and passion for the medium. So, start small, experiment, seek mentorship, and remain consistent, and you'll be well on your way to achieving radio mastery.

A STEP-BY-STEP GUIDE TO MANUAL PROGRAMMING

Baofeng radios are popular among hobbyists and professionals for their affordability and functionality. These radios are often used for two-way communication, particularly in fields like outdoor recreation, security, or emergency communication. Manual programming can be tricky for beginners, so below is a simplified step-by-step guide to help you get started.

1. **Basic Preparation:**
   - **Power On:** Turn on the radio by rotating the volume knob.
   - **Enter Frequency Mode**: Press the "VFO/MR" button to enter frequency mode.

2. **Select Frequency:**
   - **Enter Desired Frequency**: Use the keypad to input the desired frequency.
   - **Confirm Frequency**: Ensure you're operating within legal and allocated bands to avoid interfering with emergency services or other communication services.

3. **Set Transmitting Power:**
   - **Access Menu**: Press the "MENU" button.
   - **Set Power**: Input 2 or navigate to the "TXP" (Transmit Power) option, and select the appropriate power level (usually "LOW" or "HIGH").

- **Confirm Setting**: Press "MENU" again to confirm.

4. **Set CTCSS/DCS Codes (if necessary):**
   - **These codes are used for privacy. Refer to your radio's manual to set up the correct CTCSS or DCS codes.**

5. **Save to Channel:**
   - **Access Menu**: Press "MENU" again.
   - **Navigate to Memory Channel**: Input 2 7 or navigate to "MEM-CH" (Memory Channel).
   - **Select Channel**: Choose an empty channel where you wish to save the settings.
   - **Confirm Saving**: Press "MENU" to confirm saving the settings to the selected channel.

6. **Exit Programming Mode:**
   - **Press "EXIT"**: Leave the menu or programming mode by pressing the "EXIT" button.

Tips and Considerations:

- **Reference Manual**: When embarking on the task of programming your radio, it's essential to begin by consulting the reference manual provided with your specific radio model. While the fundamentals of radio programming remain consistent, the interface, options, and procedures can vary significantly from one model to another. The manual serves as your compass, guiding you through the nuances of your particular device, ensuring that you make accurate and effective adjustments.

- **Legal Compliance**: Radio frequencies are regulated by government authorities in each country, and it's imperative that you adhere to these regulations. Different countries have varying rules regarding the usage of certain frequencies, power limits, and licensing requirements. To avoid unintentional interference and legal complications, always research and comply with your country's radio frequency use regulations. This is especially crucial if you plan to operate your radio in multiple regions or while traveling internationally.

- **Battery**: Before diving into the programming process, it's crucial to ensure that your radio's battery is adequately charged. Programming a radio can be a time-consuming process, and a low battery can lead to interruptions or even data loss during programming. Ensure your radio's battery is sufficiently charged or consider using an external power source if available. This precaution will help you avoid unnecessary disruptions in communication due to a depleted battery.

- **Programming Software**: Many radio models, including those from Baofeng, offer programming software that simplifies the programming process, especially for more complex configurations. If your radio model supports it, using the manufacturer's programming software can make the task more efficient and user-friendly. Additionally, having a programming cable to connect your radio to a computer is often essential for this method. This approach allows you to program and manage a broader range of settings, including frequencies, channels, and advanced features with greater ease and precision. It also provides a visual interface that can be more intuitive than using the radio's keypad for programming tasks. If you anticipate frequently updating or customizing your radio settings, investing in a programming cable and software is a wise choice.

Incorporating these tips and considerations into your radio programming routine will not only streamline the process but also ensure that your radio operates optimally while adhering to legal requirements. By referring to the manual, complying with regulations, monitoring your battery status, and utilizing programming software when necessary, you'll enhance your radio communication experience and make the most of your equipment.

ADVANCED PROGRAMMING WITH CHIRP: TIPS AND TRICKS

CHIRP is a widely used software that provides a graphical interface for programming radios, including but not limited to Baofeng models. It supports a variety of radio brands and models, making it a popular choice among amateurs and professionals alike. Below are some advanced tips and tricks for programming with CHIRP:

1. **Getting Started:**
   - **Download and Install:**
     - Ensure that the most recent iteration of CHIRP is installed.
     - Possess the appropriate programming cable for your radio, often a USB to radio-specific adapter.
   - **Driver Installation:**
     - Install the necessary drivers for your programming cable to ensure proper communication between your computer and radio.

2. **Downloading Existing Configurations:**
   - Connect your radio to the computer using the programming cable.
   - Open CHIRP, go to the "Radio" menu, and select "Download from Radio."
   - Carefully choose the appropriate port and radio model. It is advisable to retrieve the current settings as a precautionary measure before implementing any modifications.

3. Advanced Programming Tips:

Radio enthusiasts and experts know that fine-tuning and optimizing their communication equipment frequently unlocks its actual potential. Advanced programming suggestions can improve radio communication efficiency and effectiveness beyond the basics. This section covers advanced programming choices to maximize your radio equipment.

- **Name Your Channels for Clarity**: Providing meaningful names to your channels is a fundamental yet often overlooked practice. Naming channels based on their purpose, location, or function makes them instantly recognizable, reducing confusion during use. It's a simple step that can save time and prevent errors, especially in hectic or emergency situations.

- **Duplex Settings for Repeater Use**: When using repeaters to extend your radio's range, configuring duplex settings becomes crucial. Ensure that you select the correct offset and duplex direction for the repeater you're working with. Properly configured duplex settings are essential for seamless communication in repeater-rich environments.

- **Tone Modes for Enhanced Privacy and Access**: Modern radios offer various tone modes, such as CTCSS and DCS, which can be programmed to access repeaters or create private channels. Using tone modes enhances privacy and minimizes interference, especially in crowded radio frequency environments.

- **Customize Power Levels for Different Scenarios**: Tailoring your radio's power output to the specific needs of each channel can save battery life and minimize interference. Higher power may be necessary for long-distance communication, while lower power is suitable for close-range conversations. Customizing power levels optimizes your radio's performance and extends its operational life.

- **Define Scanning Ranges for Efficiency**: Scanning is a valuable feature for monitoring multiple channels, but it can be more efficient if you define scanning ranges. By limiting the frequencies your radio scans, you can focus on the most relevant channels. This is particularly useful in crowded radio environments, where scanning a broad spectrum may lead to missed communications.

- **Program Weather Channels for Alerts**: Many radios are equipped with NOAA or other weather channels to provide critical weather alerts. By programming these channels and setting them as read-only, you prevent accidental overwrites and ensure that you receive weather updates without interruption. This is especially important in areas prone to severe weather events.

- **Set Priority Channels for Quick Access**: Priority channels are those that your radio checks more frequently during scanning. By designating channels as priority, you ensure that critical communications are not missed. This is invaluable in situations where immediate attention is required.

By employing these sophisticated programming techniques, users can optimize the capabilities of their radio communication devices, improving operational efficiency, augmenting data security, and guaranteeing dependable communication across diverse contexts. Regardless of one's experience level, these strategies can be used to optimize the performance and utility of one's radio equipment.

4. **Uploading Configurations:**

   - After making changes, save your configuration locally as a backup.

   - Go to the "Radio" menu, select "Upload to Radio," and choose the correct port and radio model.

   - Follow the on-screen instructions to complete the uploading process.

5. **Managing Multiple Configurations:**

   - The tabbed interface of CHIRP allows users to efficiently manage settings for several radios or build diverse setups for a single radio.

   - Use the export and import functionality of CSV files to facilitate the sharing of settings with others or to conveniently transition between various setups.

6. **Firmware Compatibility:**

   - Ensure your radio's firmware is compatible with the CHIRP version you're using. Check the CHIRP website for compatibility lists and known issues.

7. **Troubleshooting and Help:**

   - Refer to CHIRP's online documentation and user forums for help with troubleshooting and learning advanced features.

   - Customizing Channels and Frequencies for Various Scenarios

CUSTOMIZING CHANNELS AND FREQUENCIES FOR VARIOUS SCENARIOS

Customizing channels and frequencies on Baofeng radios can be crucial for various scenarios, such as emergency communications, outdoor recreational activities, or event coordination. Below are general steps and tips on customizing channels and frequencies for different use cases:

1. **Emergency Communications:**

- **Pre-program Frequencies**: Have local emergency frequencies, such as police, fire, and EMS, pre-programmed.

- **NOAA Weather Channels**: Program National Oceanic and Atmospheric Administration (NOAA) weather radio frequencies for real-time weather alerts and updates.

- **Repeaters**: Input frequencies of local repeaters to extend the range of communication, especially in emergency scenarios.

2. **Outdoor Recreation:**

- **GMRS/FRS Channels**: General Mobile Radio Service (GMRS) and Family Radio Service (FRS) channels are commonly used for short-distance two-way communication among small groups.

- **Local Park or Forest Service Frequencies**: Having these frequencies can be advantageous for staying informed or seeking assistance during emergencies inside national parks or reserves.

- **Simplex Channels**: Program some simplex channels for direct radio-to-radio communication without relying on repeaters.

3. **Event Coordination:**

- **Dedicated Event Channels**: Set up dedicated channels for different event coordination teams, like security, logistics, and medical services.

- **Wide/Narrow Band Setting**: Select the appropriate bandwidth setting depending on the level of traffic and the required range.

4. **General Programming Tips:**

  - **Adding Channel Names**: Name the channels descriptively for easy identification (e.g., "SECURITY", "WEATHER", "EMERGENCY").

  - **Setting Squelch Levels:** Adjust the squelch levels to filter out background noise and improve communication clarity.

  - **Privacy Codes (CTCSS/DCS):** Implement privacy codes to minimize interference from other radios on the same frequency.

  - **Power Level Adjustment:** Adjust the transmit power levels based on the necessary range and to conserve battery life.

5. **Using CHIRP for Customization:**

Utilize the CHIRP software for easier customization and programming of Baofeng radios:

  - Bulk Editing: Within the realm of radio communication management, it's often essential to make coordinated adjustments across multiple channels simultaneously. This is where the Bulk Editing feature shines. It empowers users to efficiently edit and customize a multitude of channels all at once, saving valuable time and ensuring consistent settings. Whether you're a seasoned radio technician fine-tuning a fleet of radios or a hobbyist optimizing your personal network, the Bulk Editing feature simplifies the process and ensures that all your channels are perfectly aligned with your requirements.

- Import/Export: Interoperability and data exchange are key in radio communication. The Import/Export tool makes exchanging frequencies and radio setups easy. Two major benefits come from this feature. First, it lets users input frequencies from large databases for accurate, up-to-date data without human data entry. Second, it lets users export their carefully designed radio sets for sharing or backup, ensuring they never lose crucial configurations. This feature simplifies setup and encourages radio lovers, experts, and teams to collaborate and share knowledge.

- Advanced Settings: While basic settings are essential for most radio users, there are times when more advanced control is necessary to meet specific requirements. The Advanced Settings feature offers a deeper level of customization that extends beyond what is typically available through manual programming. It provides access to parameters and adjustments that can fine-tune radio performance, optimize signal quality, and enhance security. This level of control is invaluable for professionals working in specialized fields, such as public safety or military communication, where precise configurations are mission-critical. Whether it's adjusting modulation schemes, encryption settings, or interference-reduction techniques, the Advanced Settings feature empowers users to tailor their radios to their unique needs.

6. **Responsible and Legal Use:**

   - **License Requirements**: Understand and comply with the license requirements for using certain frequencies (e.g., GMRS, ham radio).

   - **Avoid Restricted Frequencies**: Do not program or transmit on frequencies reserved for emergency services or other restricted uses.

CHAPTER 6: SIGNAL ENHANCEMENT, RANGE, AND LIMITATIONS

Baofeng radios are popular for being affordable, compact, and fairly reliable two-way radios. They are often used for amateur radio (ham radio), communication between family or team members during outdoor activities, and also by some professionals who need portable communication devices.

Signal Enhancement

Signal enhancement for Baofeng radios typically involves improving the antenna, using a repeater, or making adjustments to the settings.

1. **Antenna Upgrade**: The antenna that comes with most Baofeng radios is often not very efficient. Upgrading to a higher quality, longer, or more appropriate antenna for your use case can significantly improve the radio's signal reception and transmission range.

2. **Repeaters**: Using a repeater can extend the range of your radio. A repeater is a device that receives a signal on one frequency and retransmits it at a higher power on another frequency. In areas where there is a network of repeaters, Baofeng radios can communicate over much longer distances.

3. **Settings Adjustment**: Tweaking the settings on the radio can also help with signal enhancement. This could involve adjusting the squelch level, changing power settings, or programming the radio to access specific frequencies that might offer better reception in your area.

Range

The range of Baofeng radios can vary significantly depending on various factors:

1. **Model**: Various models exhibit varying power levels, which impact their respective ranges. Models exhibit a spectrum of power outputs, spanning from around 1 watt to 8 watts, where more wattage often corresponds to an extended range.

2. **Environment**: The extent of the range is significantly influenced by the surrounding environment. In an unobstructed open space, the range may extend up to several miles. However, in a densely populated metropolitan area characterized by many buildings or in topographically challenging terrain such as hills or forests, the range is likely to be considerably diminished.

3. **Antenna Type**: As previously stated, the use of an enhanced antenna has the potential to substantially increase the operational distance of the radio.

Limitations

Despite their popularity, Baofeng radios have several limitations:

1. **Build Quality**: Being budget-friendly devices, the build quality and durability of Baofeng radios might not match up to more expensive professional radios. They might not withstand harsh conditions or rough use as well.

2. **Legal Compliance**: Baofeng radios can sometimes transmit on frequencies that are illegal to use without a proper license. It's important to program the radios to operate only within legal frequencies and to obtain any necessary licenses, like an amateur radio license.

3. **Battery Life**: While their battery life is reasonable, heavy use, especially at high power settings, can drain the battery relatively quickly. Having spare batteries or a power bank can mitigate this issue.

4. **User Interface:** Novice users may have issues while programming and using these radios, sometimes requiring them to rely on instructional manuals or seek help from online resources. The user interface and controls of the radio systems in question may exhibit a lower degree of intuitiveness compared to their more sophisticated counterparts.

TECHNIQUES TO BOOST SIGNAL RANGE

Boosting the signal range of Baofeng radios is a common concern for many users, given that these devices are often used in various outdoor and indoor contexts where range is critical. Below are some techniques that can help increase the signal range of your Baofeng radio:

1. **Antenna Upgrade:**
   - **High-Gain Antenna**: Installing a high-gain antenna is a very influential enhancement that may be made to radio equipment. This particular antenna is specifically engineered to enhance signal strength and expand the range of communication. Various types of high-gain antennas are accessible, each specifically designed to cater to diverse purposes.
   - **Whip Antennas**: Whip antennas are often used for portable radios due to their small size and adaptable nature. The portable nature and user-friendly features of these devices provide them a significant asset to one's repertoire of communication tools, particularly in situations requiring mobility.
   - **Telescopic Antennas**: Telescopic antennas offer an adjustable length, allowing you to fine-tune your antenna to the required frequency. They are commonly used in base station setups and can be extended or retracted as needed.

- **Directional Antennas**: When precision in signal direction is crucial, directional antennas are the way to go. These antennas concentrate your radio waves in a specific direction, increasing signal range in that area while minimizing interference from other directions. They are particularly useful for point-to-point communication or long-range links.

- **Specific Use Antennas**: Your choice of antenna should align with your operational environment. For those using radios in vehicles, such as trucks or emergency response vehicles, it is advisable to consider mobile antennas designed for vehicle mounting. These antennas are built to withstand the rigors of on-the-road operation and are often magnetically or mechanically secured to the vehicle's roof or trunk. They ensure reliable communication while you're on the move, regardless of the terrain or distance.

- **Base Station Antennas**: In contrast, if your radio communication is primarily stationary, you may benefit from a base station antenna. These antennas are designed for fixed installations and are often placed atop a mast or a building. They are characterized by their height, which allows them to clear obstructions and reach farther distances. Base station antennas are commonly used in scenarios where consistent, long-range communication is essential, such as in amateur radio setups or emergency services headquarters.

2. **Use of Repeaters:**

- Repeaters are devices that receive signals from radios and retransmit them with greater power. This allows signals to cover greater distances or penetrate obstacles that would otherwise block them.

- Programming your Baofeng radio to access local repeaters can significantly extend its effective range.

3. **Power Settings Adjustment:**

- Baofeng radios often provide the capability to modify power settings. Utilizing the high power level has the potential to enhance the range; however, it is important to note that it will also result in a more rapid depletion of the battery. Modify the power configuration in accordance with the particular range requirements and the availability of power resources.

4. **Proper Placement:**

- The placement of your radio and antenna can have a significant impact on range. Elevating the antenna, keeping it vertical, and avoiding obstructions can all help improve the range.

- If indoors, try to place the radio near a window or outside wall to reduce interference from building materials.

5. **External Amplifiers:**
   - RF power amplifiers can be used to boost the transmission power of your radio. However, it's crucial to use amplifiers cautiously and legally, as increasing power can interfere with other communications and violate regulatory limits.

6. **Signal Booster Accessories:**
   - There are accessories available in the market that can act as signal boosters, helping increase the range of your radio signals without violating regulatory power limits.

7. **Optimize Device Settings:**
   - Go through your Baofeng radio's settings to optimize them for maximum range. Adjusting squelch levels, utilizing narrowband settings, and using CTSS/DCS codes can help reduce interference and improve signal clarity.

Legal Considerations:

Before implementing any of these techniques, it's essential to understand and comply with your local regulations regarding radio transmissions. In many jurisdictions, there are legal limits on transmission power, antenna types, and the frequencies that can be used without a license. Violating these regulations can result in fines, equipment confiscation, or other penalties.

ADDRESSING RANGE MISCONCEPTIONS AND LIMITATIONS

Many people who purchase Baofeng radios may have misconceptions about the range they can expect from these devices, often due to misunderstandings or unrealistic expectations set by advertising.

Misconceptions:

1. **"Maximum Range" Advertising:** Manufacturers often advertise the maximum possible range of the radios in optimal conditions, which are rarely met in real-world scenarios. For example, a claim of "up to 25 miles" may only be achievable over water or in open, flat terrain with no obstructions.
2. **Impact of Obstructions**: Users often underestimate the impact of physical obstructions like buildings, trees, hills, or even atmospheric conditions on radio signal range and clarity.

3. **Antenna Efficiency:** The stock antennas that come with Baofeng radios are not always the most efficient for every application. Users might expect far better performance without realizing an antenna upgrade is necessary.

Addressing Limitations:

Understanding the limitations of Baofeng radios helps set realistic expectations and improves user satisfaction.

1. **Power Output**: The power output of Baofeng radios is usually between 1 to 8 watts. While higher wattage can increase range, it still may not provide the extensive range some users expect, especially in challenging environments.

2. **Stock Antenna**: The included antenna is often a compromise between size and efficiency. Upgrading to a more effective antenna is one of the quickest ways to improve range and signal clarity.

3. **Environmental Factors**: Radio signals can be heavily impacted by the environment. Dense urban areas, forests, and mountainous regions can all significantly reduce range. Understanding how the environment affects radio signals is crucial for setting realistic range expectations.

4. **Frequency Limitations**: Baofeng radios operate on VHF and UHF frequencies, each having its own characteristics and limitations. VHF is generally better for outdoor environments with open space, while UHF might be more effective in urban or wooded areas.

5. **Legal Restrictions**: Regulatory bodies in different countries impose legal limits on the power output of handheld radios and the frequencies they can access. Users need to be aware of these restrictions to avoid illegal operation and potential penalties.

Overcoming Limitations:

Despite its versatility and indispensability, practical radio communication is not exempt from restrictions. In order to optimize the efficacy of one's radio communication system, it is necessary to possess the readiness and capability to surmount the obstacles that may arise. Outlined below are several ways that may be used to optimize the performance and dependability of radio communication systems:

- **Antenna Upgrade**: The antenna is an often disregarded element within the realm of radio transmission. Allocating resources towards the acquisition of a superior antenna, tailored to the precise frequency range and environmental circumstances in which it will be used, may provide significant advantages. Various types of antennas are designed and tailored to certain applications, such as facilitating short-range communication inside urban environments or enabling long-distance communication in rural or wilderness situations. The selection of an appropriate antenna has the potential to greatly improve both the range and signal quality of your radio.

- **Understand the Terrain:** The surrounding environment plays a crucial role in radio signal propagation. Factors such as buildings, trees, mountains, and bodies of water can obstruct or reflect radio waves. Understanding how your environment impacts radio signals is essential. By positioning yourself in a location that minimizes obstacles and maximizes line-of-sight to the receiver, you can improve both reception and transmission.

- **Use Repeaters:** In scenarios where you need to extend the range of communication, especially over challenging terrains or for long-distance communication, incorporating repeaters is a powerful solution. Repeaters receive a signal and retransmit it at a higher power, effectively amplifying your message and increasing the coverage area. This is particularly useful for emergency services, remote exploration, or when you need to communicate across vast distances.

- **Programming**: Properly programming your radio is crucial for optimizing the use of available frequencies and repeaters. Understanding the features and settings of your radio device will help you fine-tune its performance. You can configure it to scan for active frequencies, set squelch levels, and program emergency channels for quick access. Customizing your radio's settings to match your specific needs can enhance overall performance.

- **Licensing**: Operating a radio system legally and responsibly requires obtaining the necessary licenses and adhering to the legal framework governing radio use in your country. Licensed operators have access to additional frequencies, which can be crucial in crowded radio environments or for specialized applications. Understanding the licensing requirements and regulations not only keeps you within the bounds of the law but also ensures a more orderly and efficient use of radio frequencies.

By employing these tactics, you can circumvent numerous radio communication technology constraints. These methods will assist hobbyists, professionals, and anyone who needs radio communication for safety and coordination in maximizing its power and longevity. Your radio communication system may become more reliable and reachable by investing in the correct technology, understanding your surroundings, and following the law.

SELECTING THE PERFECT ANTENNA

Selecting the right antenna for your Baofeng radio is crucial for optimizing its performance in terms of range and signal clarity. Here's a detailed discussion on how you can select the perfect antenna:

Types of Antennas:

1. **Whip Antennas:**

   - **Rubber Duck**: The original "rubber duck" antenna that comes with the radio is compact but often inefficient. It's a compromise between size and performance.

   - **Aftermarket** Whip: Longer and potentially higher-gain than the stock antenna, an aftermarket whip can offer improved reception and transmission.

2. **Telescopic Antennas:**

   - Extendable and retractable, telescopic antennas can be adjusted for length, providing versatility in different environments. However, they can be fragile.

3. **External/Base Station Antennas:**

   - For stationary use, external antennas mounted on rooftops or high places can significantly boost your radio's range and performance.

4. **Mobile/Vehicle-Mounted Antennas:**

   - Designed for use in vehicles, these antennas are meant to enhance the performance of your radio while on the move.

5. **Directional Antennas:**

   - Also known as Yagi antennas, these focus the signal in a specific direction, providing longer range but with a narrow focus.

Factors to Consider:

1. **Frequency Band:**

   - VHF (Very High Frequency) antennas are suited for outdoor, open-space communications.

   - UHF (Ultra High Frequency) antennas are better for urban areas and through or around obstructions.

2. **Gain**:

   - The gain of an antenna is a measure of its efficiency and ability to transmit or receive signals over a given distance. Antennas with increased gain have the capability to broadcast and receive signals at extended distances, but at the expense of a reduced field of reception.

3. **Size and Portability:**

   - Smaller antennas are portable and convenient but may lack in performance. If portability is crucial, you might need to compromise on range.

4. **Durability:**

   - Depending on your usage (outdoor, rough environments), you may need an antenna that is robust and can withstand various conditions.

5. **Connector Type:**

   - It is important to verify that the antenna has the appropriate connection type that is compatible with the specific Baofeng radio model in question.

6. **Brand Reputation:**

   - Consider antennas from reputable brands, as these tend to offer better performance and durability.

Recommendations:

- **For General Use**: In the realm of basic radio communication applications, opting for an aftermarket whip antenna with a modest level of gain is often seen as a flexible selection. These antennas demonstrate versatility across many situations and provide a favorable balance between dimensions, signal amplification, and operational effectiveness. The use of either a handheld radio or a portable transceiver warrants consideration of a whip antenna as a pragmatic choice, as it offers dependable communication capabilities across a diverse array of circumstances. The radio's user-friendly interface and convenient mobility make it a very favorable option for those seeking adaptability in their radio configurations.

- **For Vehicle Use**: It is strongly advised to use a magnetic-mount mobile antenna that is specially suited for the intended frequency band while using radio equipment in a car. The design of these antennas is optimized for best performance when installed on the roof or trunk of a car. The magnetic base provides a reliable and stable connection, and is available in a range of forms and lengths to fit diverse vehicle types. Regardless of one's career or personal interests, the use of a vehicle-mounted antenna guarantees the ability to sustain consistent and dependable contact while in transit.

- **For Base Station**: For those establishing a fixed base station, particularly for amateur radio or emergency communications, a high-gain external antenna is the way to go. These antennas are designed for long-range communication and are typically installed on rooftops, towers, or masts. They can significantly enhance your signal's reach, making them suitable for scenarios where extended coverage is necessary. If you have specific areas or directions you need to target, consider a directional antenna, such as a Yagi or log-periodic antenna, which can focus your signal in a specific direction for more precise communication.

- **For Varied Environments**: In some cases, your radio communication needs may span diverse environments and scenarios, from urban areas to remote wilderness. To ensure optimal performance across these settings, it's advisable to maintain a diverse array of antennas tailored to specific applications. For instance, having both a compact whip antenna for portability and a high-gain directional antenna for long-distance communication allows you to adapt to different situations effectively. A well-prepared radio operator will have an arsenal of antennas at their disposal, ensuring that they are ready for any communication challenge that may arise.

The appropriate selection of an antenna is of utmost importance in order to achieve the intended range and signal quality in the realm of radio communication. By taking into account these suggestions and customizing your selection of antennas to suit your particular requirements, you can optimize the efficiency of your radio devices across a range of environments and use cases. Regardless of one's level of experience as a radio user, the selection of an appropriate antenna plays a crucial role in facilitating effective and dependable communication.

POWER SYSTEMS

The power system of Baofeng radios primarily consists of the power source (usually a battery), power management, and the power output used for transmission. Below, you'll find details about the various elements of the power systems within these radios:

1. **Battery System:**

   - **Li-ion Batteries**: Most Baofeng radios use rechargeable Lithium-ion (Li-ion) batteries. Li-ion batteries are preferred for their lightweight, high energy density, and ability to handle numerous charge cycles.

   - **Battery Capacity**: The battery capacity for Baofeng radios typically ranges from 1500mAh to 2800mAh. Higher capacity batteries offer longer operation time but might be bulkier.

   - **Battery Life**: The life of a Baofeng radio battery on a full charge will depend on the usage pattern. Continuous transmission drains the battery faster than receiving, and standby mode consumes the least power. On average, you might expect anywhere from 12 to 20 hours of use on a full charge, depending on the model and usage.

   - **Replacement and Backup**: Always have spare batteries or a backup power source, especially if you plan to use the radio extensively without access to charging facilities.

2. **Charging System:**

- **Charging Base**: Many Baofeng models come with a drop-in charging base, making it convenient to charge the radio. Ensure the radio is properly seated in the base for effective charging.

- **USB Charging Option**: Some models offer USB charging capabilities, providing flexibility in charging the radio from different power sources, including power banks and car chargers.

- **Charging Time**: Charging time varies, usually taking around 3 to 5 hours to fully charge a depleted battery. It's essential to monitor the charging process to avoid overcharging, which might reduce battery life.

3. **Power Output:**

- **Adjustable Power Levels**: Baofeng radios often come with adjustable transmission power settings, typically ranging from 1 to 8 watts. Higher power settings will increase the transmission range but also drain the battery faster.

- **Power Consumption**: The power consumption in standby mode, reception mode, and transmission mode varies significantly, with transmission consuming the most power.

4. **Power Management:**

- **Battery Saver Mode**: Some radios have battery-saving features that reduce power consumption when the radio is inactive or in standby mode.

- **Voltage Display**: Radios often have a display showing the current battery voltage, helping users monitor the battery level and plan for recharging or battery replacement. level and plan for recharging or battery replacement.

Tips for Effective Power Management:

The optimization of efficiency and durability for a Baofeng radio, or any other portable electronic equipment, is contingent upon the use of appropriate power management strategies. Power management encompasses more than just extending the lifespan of a battery; it also entails preparing equipment for optimal functionality at critical moments. Presented below are a set of practical recommendations aimed at optimizing the use of the power supply for your radio device:

- **Regular Charging**: Regular charging is a vital component of efficient power management, despite its seeming simplicity. It is advisable to develop a routine of frequently recharging batteries, even if they have not been fully discharged. Contemporary rechargeable batteries, such as those used in Baofeng radios, are not susceptible to the phenomenon known as "memory effect" that was seen in previous battery technologies. Consequently, there is no detrimental impact associated with regularly replenishing the battery charge. This method guarantees the continuous preparedness of your radio, thus mitigating the potential hazard of its power depletion at a crucial juncture.

- **Avoid Overcharging**: While it is important to engage in frequent charging, it is equally vital to refrain from leaving batteries on the charger once they have reached a full charge. Excessive charging has the potential to result in a gradual decline in both battery longevity and operational effectiveness. The majority of contemporary chargers and electronic devices are engineered with built-in mechanisms to mitigate the risk of overcharging. Nevertheless, it is advisable to disconnect the power supply from your radio after it has reached full charge, particularly if immediate use is not anticipated.

- **Battery Storage**: In the event that one anticipates a prolonged period of inactivity for their Baofeng radio, it is advisable to remove the battery. One possible strategy to mitigate potential harm caused by leakage or corrosion over time is to store the radio without the batteries. Furthermore, it is advisable to keep the battery in a cool and dry environment, avoiding exposure to direct sunlight and severe temperatures, since these factors may have a detrimental impact on the longevity of the battery.

- **Invest in Quality Accessories**: The acquisition of high-quality accessories for replacement or backup batteries and chargers might provide substantial benefits. Choose accessories that are particularly tailored for your Baofeng radio model and are known for their dependable performance. The use of high-quality accessories not only enhances the optimal performance of your radio but also mitigates the potential hazards associated with damage or compatibility discrepancies.

By following these tips for effective power management, you can extend the life of your Baofeng radio's battery, maintain its efficiency, and be confident that it will reliably serve your communication needs. Whether you use your radio for recreational activities or as a vital tool in your profession, ensuring that it's always ready for action is key to a seamless and uninterrupted communication experience.

POWER CONNECTORS

The power connectors on Baofeng radios primarily pertain to charging interfaces and external power sources. The following sections discuss the most common types of power connectors associated with Baofeng radios, along with considerations for their use:

1. **Battery Charging Connectors:**

   - **Drop-In Charging Dock**: Many Baofeng radios come with a charging dock or cradle designed to hold the radio or its battery for charging. The radio or battery is placed into this dock, aligning with the embedded connectors. These docks usually connect to power sources through AC adapters.

   - **Direct Plug-In Connector**: Some radios have direct plug-in connectors on the device or the battery itself, allowing you to connect an AC adapter or charger directly without needing a charging dock.

2. **USB Charging Ports:**

   - **Micro-USB or USB-C Ports**: Some newer or upgraded Baofeng radio models may feature Micro-USB or USB-C ports, making it easier to charge the radio using standard USB cables and adapters. This feature adds convenience, as these cables are widely available and might already be in use for other devices.

3. **External Power Accessories:**

   - **Battery Eliminators**: Battery eliminators are accessories that replace the radio's battery, allowing the radio to be powered directly from a vehicle's electrical system or another external power source. These accessories typically have connectors that fit into the car's cigarette lighter or 12V power socket.

4. **Connector Polarity and Specifications:**

   - When using external power sources, chargers, or adapters, it's critical to ensure that the connector's polarity and voltage specifications match the radio's requirements. Using incompatible connectors or power sources can damage the radio and void the warranty.

Considerations for Power Connectors:

   - **Compatibility**: Always use chargers, docks, and accessories designed for or compatible with your specific Baofeng radio model. Compatibility ensures that the connectors fit securely and supply the correct voltage and current.

   - **Care and Maintenance**: Connectors should be kept clean and dry. Periodically inspect connectors for damage, wear, or corrosion, and replace accessories if needed.

   - **Safety Precautions**: Only use power sources that meet the safety standards and regulations applicable in your region. Avoid using damaged or frayed cables, as these could pose a risk of electric shock or fire.

- **Original Equipment Manufacturer (OEM) Accessories**: While third-party accessories can be cost-effective, using OEM accessories usually provides a better guarantee of compatibility and safety.

RADIO BAND GUIDE

Baofeng radios are dual-band radios capable of operating on Very High Frequency (VHF) and Ultra High Frequency (UHF) bands. The specific frequencies within these bands that a Baofeng radio can access vary depending on the model, but most are designed to cover the following ranges:

1. **VHF (Very High Frequency) Band:**

   - **Frequency Range**: Typically ranges from 136 to 174 MHz.

   - **Common Uses:**

     - Marine communication.

     - Amateur (ham) radio - 2-meter band (144 to 148 MHz).

     - Aviation and air traffic control communication.

     - Some public safety and emergency services.

   - **Characteristics:**

     - Better suited for open or outdoor areas with less obstruction.

     - In general, Very High Frequency (VHF) exhibits longer wavelengths, enabling the transmission of signals over greater distances while requiring relatively lower power levels.

     - A decrease in frequency corresponds to a reduced probability of absorption by structures and topography.

2. **UHF (Ultra High Frequency) Band:**

   - **Frequency Range**: Usually ranges from 400 to 520 MHz.

   - **Common Uses:**

     - Business band.

     - GMRS (General Mobile Radio Service).

     - FRS (Family Radio Service).

     - Amateur (ham) radio - 70cm band (420 to 450 MHz).

   - **Characteristics:**

     - More suitable for indoor use or areas with obstructions like buildings, as UHF signals can penetrate through concrete and steel more effectively.

     - Higher frequency means shorter wavelength, making UHF more effective for communication in dense urban areas.

Understanding Bands for Amateur Radio:

For those interested in using Baofeng radios for amateur (ham) radio purposes:

- **2 Meter Band (VHF):** This band is popular among amateur radio operators. It is primarily used for local communication and is available to Technician class (or higher) licensees.
- **70 Centimeter Band (UHF):** This is another popular band for amateur radio, especially in cities where repeaters might re-broadcast signals across a wider area.

Licensing and Legal Considerations:

The realm of radio communication, with its wide array of applications and frequency bands, is not only a powerful tool for connectivity but also a regulated domain governed by licensing requirements and legal considerations. In our ever-connected world, understanding the nuances of licensing and adhering to the legal framework surrounding radio communication is paramount for ensuring both responsible use and compliance with local regulations.

- **Licensing:** Individuals should be aware that transmission activities on certain frequencies may require the possession of a license, such as an amateur radio license or a General Mobile Radio Service (GMRS) license within the jurisdiction of the United States.
- **Regulations:** Each country has different regulations concerning radio frequencies and bands. Ensure you are familiar with and adhere to the local regulations and laws regarding radio transmission to avoid penalties.

In conclusion, licensing and legal considerations are integral facets of the radio communication landscape. By familiarizing yourself with licensing requirements and local regulations, you not only demonstrate your commitment to responsible radio use but also ensure that you operate within the bounds of the law, contributing to the overall reliability and effectiveness of radio communication systems.

Programming and Channel Setup:

- Baofeng radios are programmable, allowing users to set up channels for specific frequencies and bands. Understanding the bands and their appropriate uses will guide you in programming your radio effectively.
- Use programming software like CHIRP to facilitate the process of programming and managing the channels on your Baofeng radio.

A FIELD EPISODE: THE RESILIENCE OF RELIABLE COMMUNICATION

Dear radio adventurers,

During one of my excursions in the dense forests of the north, I faced an unexpected challenge.

A storm had darkened the sky and cut off cellular communications. At that moment, I relied on my Baofeng radio, a tool on which I had based much of my research and experience. Thanks to the "advanced scanning" technique I delved into in the book, I quickly identified an active local emergency frequency.

Using the emergency call protocol I explained in the initial chapters, I managed to connect with a rescue group and coordinate with them until normal communications were restored. This experience underscored the importance of thorough preparation and the versatility of the Baofeng radio in critical situations.

We are now travel companions in "The Baofeng Radio Bible", and I hope the journey thus far has revealed new horizons and provided valuable tools and insights. As you read from the feedback of Cip, this book is more than just a guide.

If you find this guide valuable, please consider sharing your experience with others who might benefit from it. Your support is not only appreciated but also instrumental in fostering a community of well-prepared individuals.

Now, I've already mentioned how challenging and vitally important it is for the success of the book to receive readers' consent.

Often taken for granted, know that a feedback on Amazon can light the way for other curious readers like you. So, I want to help make your evaluation as objective and honest as possible.

When considering writing your feedback on Amazon, think about how detailed preparation can make a difference in times of need. You might focus on aspects like:

- **Depth of Coverage:** Did the guide cover all the aspects necessary to make you feel prepared?

- **Practical Instructions:** How did you find the usefulness of the techniques in a real-world context? Did I provide you with the necessary confidence to handle emergency communication?

- **Educational Impact:** Which teachings or chapters equipped you best for your radio adventures?

Do it now and leave an honest feedback; go to the ORDERS section of your Amazon account and click on the "Write a review for the product" button , or scan this QR code.

Additionally, if you feel that certain aspects of the book could be improved and/or to continuously enrich this resource to meet the needs of anyone who may find themselves facing similar situations, I invite you to write to me with any suggestions or simply share your experiences with me. My inbox is always open: **info@survivalhorizon.com**.

Together towards a future of safer and more reliable communications,

Maxwell Cipher

CHAPTER 7: BAOFENG IN EMERGENCIES: A SURVIVAL GUIDE

Baofeng radios are popular for their affordability and utility in various emergency situations. These two-way radios are commonly used for communication over short to medium distances, proving particularly handy during disasters or emergencies where cellular networks might be down or overloaded.

Survival Guide: Using Baofeng in Emergencies

1. **Choosing the Right Model:**

Baofeng offers a range of distinct models, each with unique attributes and capabilities. When making a decision, it is important to carefully evaluate and choose the model that aligns most effectively with one's requirements, taking into account factors such as battery life, range, and the number of accessible channels. The Baofeng UV-5R and BF-F8HP models are often favored for survival applications due to their notable adaptability and robust performance capabilities.

2. **Programming:**
   - Manual Programming: You can manually input frequencies into the radio. Familiarize yourself with the keypad and the steps needed to enter frequencies, set the squelch level, and save channels.
   - Computer Programming: Using programming cables and software like CHIRP, you can easily program your radio with various frequencies and settings.

3. **Understanding Frequencies:**
   - Baofeng radios operate on VHF (Very High Frequency) and UHF (Ultra High Frequency). Understand the legal implications of transmitting on certain frequencies, as some are reserved for specific services.
   - Familiarize yourself with the frequencies that local emergency services, ham radio operators, and weather stations use in your area.

4. **Legal Compliance:**
   - Transmitting on certain frequencies requires a license (e.g., an amateur radio license in the U.S.). Understand and comply with the laws and regulations to avoid penalties.

5. **Battery Management:**
   - Baofeng radios usually come with rechargeable batteries. Have backup power sources available, like solar chargers or extra battery packs.
   - Conserve battery by turning off unnecessary features and reducing the transmit power when possible.

6. **Preparation and Practice:**

- Learn to operate the radio proficiently before an emergency arises. Practice using its various features and functions.
- Label programmed channels for easy identification and access during emergencies.

7. **Maintenance:**
   - Store your radio in a cool, dry place and protect it from extreme temperatures and moisture.
   - Regularly check the condition of the antenna, battery, and other components.

8. **Communication Protocols:**
   - Strive to acquire and comprehend fundamental communication protocols and etiquettes, including the use of clear language, the practice of concise transmissions, and the correct identification of oneself.

Using Baofeng During Emergencies:

1. **Emergency Communication:**
   - In case of disasters, use Baofeng radios for short-range communication with family, neighbors, or emergency responders.
   - Monitor emergency frequencies for important updates and information.

2. **Situational Awareness:**
   - Stay informed about the situation in your area by tuning into weather, news, or emergency services channels.

3. **Search and Rescue:**
   - Baofeng radios can be crucial tools for coordinating search and rescue efforts, especially in areas with no cellular reception.

CRAFTING AN EFFECTIVE EMERGENCY COMMUNICATION PLAN

1. **Identification and Assessment:**
   - Identify potential risks and emergencies in your area (e.g., hurricanes, wildfires, floods, etc.).
   - Assess your communication needs based on your environment, family size, and community infrastructure.

2. **Selecting Your Equipment:**
   - Choose a suitable Baofeng model.
   - Invest in essential accessories like antennas for extended range, extra batteries, and protective cases.

3. **Programming & Familiarization:**
   - Pre-program important frequencies into the radio.
   - Familiarize yourself and all users with the equipment, ensuring everyone knows the basic operations.

4. **Licensing & Legalities:**
   - Acquire the necessary licenses for legal transmission (e.g., Amateur Radio License).
   - Understand the legalities and restrictions associated with using certain frequencies.

5. **Establishing Communication Protocols:**
   - Develop a set of standardized communication procedures and codes for efficient transmission.
   - Educate all family members or team participants about these protocols.

6. **Regular Training & Drills:**
   - Conduct regular training sessions and drills with all potential users to reinforce learning and practice emergency communication.
   - Drills should simulate real-life emergency situations to improve responsiveness and familiarity with protocols.

7. **Designating Emergency Contacts:**
   - Have a list of emergency contacts and important frequencies readily available.
   - Assign specific roles and responsibilities to each member of your group or family, ensuring everyone knows their part in an emergency.

8. **Battery & Power Management:**
   - Plan for long-term power outages. Have a strategy for charging or replacing batteries, such as having solar chargers, hand-crank generators, or ample spare batteries.

9. **Safe Storage & Maintenance:**
   - Store your Baofeng radio and accessories in an easily accessible yet secure location.
   - Regularly check and maintain your equipment to ensure it's in working order when needed.

10. **Continuous Improvement:**
    - After every drill or actual emergency, review and assess the effectiveness of your communication plan.
    - Make necessary adjustments and improvements based on lessons learned and feedback from participants.

Emergency Communication Plan Example:

1. **Pre-Emergency Preparation:**
   - Clearly list out roles and responsibilities for each participant.
   - Have a laminated card with important frequencies, emergency codes, and protocols attached to each radio.

2. **During an Emergency:**
   - Follow predefined roles and communication protocols.
   - Use coded language if necessary for privacy, but keep communication clear and concise.

- Continually monitor designated emergency frequencies for important updates and instructions from local authorities.

3. **Post-Emergency Evaluation:**
   - Once the situation is safe and secure, gather participants for a debriefing.
   - Discuss what worked well and identify areas needing improvement in the communication plan or equipment usage.

ESSENTIAL FREQUENCIES FOR CRISIS SITUATIONS

In a crisis or emergency, communication is crucial. Having access to the right frequencies can make a significant difference. When using Baofeng radios, or any two-way radios, it's important to understand and respect the regulations pertaining to the use of radio frequencies in your country, as unauthorized transmission on certain frequencies is illegal.

Essential Frequencies for Crisis Situations:

1. **National Weather Service (NWS)/NOAA Weather Radio (NWR):**
   - **Frequency Range**: 162.400 MHz to 162.550 MHz
   - These frequencies provide uninterrupted dissemination of up-to-date weather data and prompt notifications specific to your geographical location.

2. **Family Radio Service (FRS):**
   - **Frequency Range**: 462.5625 MHz to 467.7125 MHz
   - FRS is a set of frequencies that don't require a license to transmit in the United States. They're useful for short-distance communication between family members or small groups.

3. **General Mobile Radio Service (GMRS):**
   - **Frequency Range**: 462.550 MHz to 467.725 MHz
   - GMRS frequencies provide superior coverage compared to FRS frequencies. Nevertheless, in the United States, obtaining a license is required for transmitting on certain particular frequencies.

4. **Amateur Radio (Ham) Frequencies:**
   - **2 Meter Band**: 144 MHz to 148 MHz
   - **70 Centimeter Band**: 420 MHz to 450 MHz
   - These bands are often used by those who engage in amateur radio operations. They possess commendable capabilities for facilitating communication, both within close proximity and at long distances. In order to broadcast on these frequency bands, it is necessary to possess a valid license.

5. **Multi-Use Radio Service (MURS):**

- **Frequency Range**: 151.820 MHz to 154.600 MHz

- MURS frequencies are available for public use without the need for a license, making them a practical option for engaging in short-range, bidirectional communication.

6. **Emergency Frequencies:**

- **CB Radio Channel 9**: 27.065 MHz

- National Simplex Calling Frequency for 2m: 146.520 MHz

- These are universally recognized for emergency communication. While CB radio isn't accessible on Baofeng, the 2-meter ham frequency is available for licensed users.

Programming and Usage Tips:

- **Pre-program Frequencies**: One of the fundamental steps in using your Baofeng radio effectively is to pre-program essential frequencies. By programming frequently used channels, you save time and ensure that you can quickly access the communication channels you need. Whether you're an amateur radio enthusiast, a security professional, or an outdoor enthusiast, this practice ensures you're always prepared.

- **Respect Licensing Requirements**: Radio frequencies are a shared resource, and many require a license for transmission. It's imperative to understand and adhere to these licensing requirements. Operating on frequencies without the necessary license can lead to legal repercussions. Moreover, using frequencies open to the public without a license helps maintain an organized and interference-free radio environment for everyone.

- **Monitor First**: A crucial aspect of responsible radio communication is to listen in on a frequency before transmitting. This practice, often referred to as "listening before talking," allows you to get a sense of the ongoing communication. By doing so, you can avoid interrupting important discussions and maintain good radio etiquette. It's especially important in emergency situations where clear and uninterrupted communication is vital.

- **Understand the Protocol**: Different frequencies or bands might have their own set of protocols or etiquettes. For example, amateur radio operators follow specific procedures, and emergency services employ their own communication protocols. Understanding these protocols is essential to effective communication. It ensures that your messages are conveyed accurately and in a way that is comprehensible to others on the same frequency.

- **Emergency Broadcasts:** In various regions, local emergency frequencies are designated for disaster alerts and critical communication during emergencies. These frequencies are often monitored by local authorities and first responders. As a responsible radio user, you should be aware of these channels and their intended usage. During emergencies, refrain from transmitting on these channels unless you have a genuine emergency situation to report. Listening to these channels can provide valuable information during times of crisis.

Incorporating these programming and usage tips into your Baofeng radio operations not only enhances your communication experience but also contributes to a more organized and respectful radio community. By being diligent about frequency programming, adhering to licensing requirements, practicing proper monitoring and protocol, and understanding the significance of emergency broadcast channels, you not only improve your own radio communication skills but also ensure the reliability and integrity of the entire radio communication system.

Responsible and Legal Use:

- Avoid transmitting on frequencies allocated for emergency services, government agencies, or military use unless you are authorized.

- Stay informed about the regulations concerning radio transmission in your country to avoid legal consequences.

Understanding how to use frequencies responsibly and effectively requires ongoing learning and practice. Consider joining a local amateur radio club to improve your skills and knowledge about handling radios in emergencies.

Before relying on any frequencies in an emergency, it's critical to verify the information, as the allocation of frequencies can vary significantly from country to country and even within regions in the same country. My knowledge has a cutoff in 2022, and rules, regulations, or frequency allocations might have changed after this time, so always refer to the most current and authoritative sources for accurate and legal guidance.

REAL-LIFE TESTIMONIES: BAOFENG IN ACTION

Baofeng radios, due to their affordability and accessibility, have been used in various real-life emergency situations by individuals and emergency responder groups alike. While specific personal testimonies aren't available due to privacy concerns and my inability to browse, general anecdotes and use cases can be discussed to illustrate the value of these radios in emergencies. Below are some general scenarios where Baofeng radios have proven valuable:

1. **Search and Rescue Operations:**
    *   **Scenario**: In wilderness or mountainous areas where cell reception is unreliable, Baofeng radios have been used for communication between search and rescue teams. These radios facilitate coordination and strategy implementation, especially in terrain where line-of-sight communication is essential.
    *   **Testimony**: Individual rescuers and volunteers often share stories about how these radios have been indispensable in organizing and conducting rescue missions efficiently.

2. **Natural Disaster Response:**
    *   **Scenario**: During hurricanes, floods, or earthquakes, when conventional communication infrastructure may be damaged, Baofeng radios provide a reliable alternative. Civilians and responders use them to communicate critical information, call for help, or coordinate relief efforts.
    *   **Testimony**: There have been anecdotal accounts from community members and emergency responders about using Baofeng radios to stay connected during crises, aiding in both evacuation processes and immediate post-disaster scenarios.

3. **Outdoor and Adventure Activities:**
    *   **Scenario**: Adventure enthusiasts, hikers, and campers often carry Baofeng radios as part of their safety gear. In situations where individuals get lost or need assistance, these radios serve as vital communication tools when there's no cell service.
    *   **Testimony**: Various outdoor enthusiasts recount situations where the radio was crucial in contacting fellow group members or emergency services during unexpected events in remote locations.

4. **Community Emergency Preparedness:**
    *   **Scenario**: Communities prone to natural disasters often form emergency preparedness groups. These groups use Baofeng radios to create communication networks, enabling quick dissemination of information and alerts among neighbors.

- **Testimony**: Residents in various communities have reported that having a pre-established radio network significantly improved their collective response and coordination during emergencies.

5. **Event Coordination and Security:**
   - **Scenario**: During large public events, security and volunteer staff use Baofeng radios for instant communication. In emergency situations or medical crises during the event, radios facilitate swift coordination for help and evacuation.
   - **Testimony**: Event organizers and security personnel often testify to the usefulness of these radios in ensuring smooth communication, which is crucial for the safety and success of the event.

Note on Responsible and Legal Use:

It is crucial to highlight that while Baofeng radios can be powerful tools, users must adhere to legal and ethical standards when operating them. Unauthorized transmission on certain frequencies is illegal and can interfere with critical communication channels used by emergency services, potentially leading to dangerous situations and legal consequences.

For specific and detailed personal testimonies, consider researching through online forums, social media platforms, and community discussion groups related to emergency preparedness, amateur radio, and Baofeng users. These platforms often have individuals sharing their personal experiences and stories about using Baofeng radios in various emergency scenarios. Always cross-verify any information you obtain from these sources, as the accuracy and reliability of personal testimonies can vary.

CHAPTER 8: TROUBLESHOOTING, MAINTENANCE, AND BATTERY ISSUES

Baofeng radios, often utilized by radio amateurs and enthusiasts for communication, are affordable, portable devices known for their reliability and ease of use. Despite these advantages, users may still encounter some common problems, including issues with troubleshooting, maintenance, and batteries. Below are some details addressing each concern:

Troubleshooting

Radio communication is a critical tool in various industries, including public safety, transportation, and among amateur radio enthusiasts. While radio devices are generally reliable, issues can occasionally arise. Understanding how to troubleshoot these problems is essential to ensure seamless communication. Here are some common issues and troubleshooting steps to address them:

1. **No Power:**

Your communication might be disrupted by a dead radio. Check the power supply first. The battery should be completely charged or replaced. Make sure the battery is properly placed and connected. Corrosion or damage to battery terminals may reduce power flow. In case of corrosion, clean the terminals using battery terminal cleaner or a mixture of water and baking soda.

2. **Poor Reception:**

Weak or no signal can be frustrating, especially when you rely on your radio for communication. If you're experiencing reception issues, consider the following:

- **Adjust the antenna**: Sometimes, a slight change in the antenna's position can significantly improve signal strength.
- **Change your location**: If you're in a building or an area with many obstructions, moving to a more open space can often improve reception.
- **Minimize electronic interference**: Electronic devices such as computers, routers, or other radios can interfere with reception. Ensure that your radio is away from potential sources of interference.

3. Audio Problems:

If you encounter distorted or unclear audio, addressing this issue is crucial for effective communication. Try the following steps:

- **Adjust the volume**: Sometimes, distorted audio is a result of an inappropriate volume setting. Lower or raise the volume to a more suitable level.
- **Seek a better signal**: Poor audio quality can often be attributed to a weak signal. Moving to an area with a stronger signal can resolve this problem.
- **Check the speaker**: If the speaker is damaged or malfunctioning, it can lead to audio issues. In such cases, consider replacing the speaker with a compatible one to restore clear sound quality.

4. Programming Issues:

Radio programming can be a complex task, and users may encounter difficulties. Here are some tips to tackle programming issues:

- **Consult the manual**: Most radios come with user manuals that provide step-by-step instructions for programming. Review the manual carefully to ensure you follow the correct procedure.
- **Use programming software**: Many radios can be programmed using specialized software. Install the software and connect your radio to a computer for easier programming.
- **Seek assistance**: If you're still facing issues, don't hesitate to seek help from online forums or experienced users. They may provide valuable insights and solutions to your specific programming problems.

By addressing these common radio communication issues, you can maintain the reliability and effectiveness of your radio equipment, ensuring that you stay connected when it matters most. Remember that regular maintenance and following the manufacturer's guidelines for your specific radio model are essential for trouble-free communication.

Maintenance

Maintenance is a crucial aspect of ensuring the longevity and optimal performance of your radio communication equipment. Neglecting proper maintenance can lead to diminished functionality and even device failure. Here are some extended guidelines on how to effectively maintain your radio communication gear:

- **Cleaning**: Regularly wipe down the radio with a soft, dry cloth to remove dust, dirt, and grime. This simple step helps keep your device looking clean and ensures that buttons, knobs, and the screen remain responsive. Avoid using harsh chemicals or abrasive materials, as they can potentially damage the plastic components and screen. Instead, opt for specialized electronic cleaning solutions if necessary.

- **Storage:** Ensuring appropriate storage is crucial for the protection and preservation of radio equipment. It is advisable to store the radio in a location that is cool, dry, and shielded from direct exposure to sunlight. The device's longevity may be compromised and its functionality impaired as a result of exposure to extreme temperatures, whether excessively hot or cold. Furthermore, it is important to safeguard your radio equipment from the detrimental effects of moisture or humidity, as these environmental factors have the potential to induce corrosion in the internal components, therefore jeopardizing the overall functioning of the device. It is advisable to contemplate the allocation of resources towards the acquisition of a protective carrying case, particularly if the radio device is regularly subjected to transportation.

- **Antenna Care:** The antenna is a critical part of your radio's performance. Handle it gently and avoid any harsh bending or pulling, as this can lead to damage and result in decreased signal reception. Regularly inspect the antenna for signs of wear and tear, such as frayed wires or loose connections. Ensure that it is securely attached to the radio to maintain optimal signal transmission.

- **Software Updates:** In order to optimize the functionality and performance of your radio, it is important to stay current with software updates. To get information on available updates, it is recommended to consult the manufacturer's official website or refer to the user manual provided with the device. The installation of these updates has the potential to enhance performance and perhaps resolve security vulnerabilities or issues that have been identified subsequent to the manufacturing of your device.

- **Accessories Care:** In addition to maintaining the radio itself, it's equally important to regularly check and maintain any accessories you use with your radio. Items like microphones, earpieces, and programming cables can suffer from wear and tear over time. Inspect these accessories for signs of damage, such as frayed wires, loose connections, or worn-out ear cushions. Keeping these accessories in good condition ensures that they function reliably when you need them most and extends their lifespan.

By following these comprehensive maintenance guidelines, you can enjoy long-lasting and trouble-free use of your radio communication equipment. Proper care and attention to detail not only preserve your investment but also ensure that your radio is always ready to keep you connected and informed in a variety of situations.

Battery Issues

- **Short Battery Life:** The radio's battery life might diminish over time. To prolong it, turn off unnecessary features, reduce the backlight brightness, and use power-saving modes when available.
- **Battery Won't Charge:** In the event that the battery fails to charge, potential sources of the issue may include the charger, charging cable, or the battery itself. Conduct independent tests on each component to find the underlying problem.

- **Replacement Batteries**: It is vital to use batteries that are explicitly developed for the designated Baofeng radio model while undertaking the task of battery replacement. The use of batteries that are not compatible with the radio might lead to detrimental effects such as damage to the device or suboptimal performance.

- **Battery Maintenance**: Store batteries in a cool, dry place. Avoid exposing them to high temperatures, as heat can reduce their lifespan. Also, consider using a smart charger that prevents overcharging, which can also extend battery life.

COMPREHENSIVE GUIDE TO COMMON ISSUES

Baofeng radios are popular for their affordability and utility in various applications. Like all electronic devices, they have their fair share of common issues. Understanding these problems can help users troubleshoot and maintain their radios effectively. Here's a comprehensive guide to some common issues:

1. **Power and Battery Issues:**
   - **No Power:** Ensure that the battery is charged and properly seated in the radio. The terminals should be clean, with no signs of corrosion or damage.
   - **Short Battery Life:** The anticipated lifespan of a Baofeng radio battery typically ranges from 8 to 12 hours. In the event of a significant decrease in battery life, it is advisable to explore potential solutions such as adjusting the screen brightness, deactivating superfluous functionalities, or contemplating the replacement of the battery.
   - **Battery Won't Charge**: If the battery isn't charging, check the charger and charging cable for damages. If both are functional, the battery itself might be defective.

2. **Audio and Reception Issues:**
   - **Poor Reception**: Weak signals might be due to physical obstructions, distance from the signal source, or antenna issues. Adjust the antenna, change the location, or consider getting a high-gain antenna for better reception.
   - **Distorted Audio**: Distorted or unclear audio may result from volume misadjustment, speaker damage, or poor reception. Adjust the volume, move to a clearer reception area, or consider replacing the speaker.

3. **Programming and Software Issues:**
   - **Programming Difficulties**: Users often have challenges programming the radios for the first time. Ensure you are using the correct programming cable and software compatible with your Baofeng model. Reading the manual or watching online tutorials can also help.

- **Software Compatibility**: If the programming software isn't compatible with your operating system, consider running it in compatibility mode or using a different computer.

4. **Hardware and Accessory Issues:**
   - **Defective Microphone or Speaker:** In the event of a malfunctioning internal microphone or speaker, using an external microphone or earphone might serve as a temporary remedy. For a lasting resolution, it is advisable to contemplate the replacement of the defective components.
   - **Malfunctioning Buttons**: Buttons can wear out or malfunction over time. Cleaning around the buttons may help, but if the issue persists, you may need professional repair.

5. **Maintenance and Care:**
   - **Regular Cleaning**: To keep your radio equipment in pristine condition, it's advisable to perform regular cleaning. Use a soft, dry cloth or a slightly damp one to wipe down the radio body. Avoid using abrasive or harsh chemicals, as they can damage the exterior and compromise the device's overall integrity. By removing dust, dirt, and grime, you not only maintain the device's appearance but also prevent potential damage that can occur when foreign particles infiltrate internal components.
   - **Protective Case Usage**: For added protection, consider investing in a protective case designed specifically for your radio model. These cases are engineered to safeguard your device from accidental drops, impacts, dust, and moisture. A protective case can serve as a shield, preventing damage that might otherwise occur during outdoor activities or in rugged work environments. It's a simple yet effective way to extend the life of your equipment and ensure it remains in optimal working condition.
   - **Antenna Care:** Radio antennas are crucial for signal transmission and reception. They must be handled carefully. Avoid bending, twisting, or rough handling of the antenna to prevent damage and performance issues. Consider replacing the antenna if it has malformations, cracks, or difficulties. Damaged antennas reduce signal quality and radio capability. To ensure reliable connectivity, check the antenna for wear and tear regularly.

6. **Firmware and Updates:**
   - **Outdated Firmware**: Ensure your radio is running the latest firmware. If not, consider updating it for enhanced functionality and bug fixes.

7. **Operational Errors:**
   - **Incorrect Frequency Entry**: Selecting the right frequency channel is crucial for radio transmission. In crucial cases, frequency entering errors may disrupt communication and endanger safety. To verify

channel match, double-check the frequency entered. Misinterpretation of channel numbers, typos, and external interference may cause this issue. Human error may lead to frequency input errors, thus training and education are crucial. Use of radio equipment with user-friendly interfaces and error-checking algorithms may further reduce this sort of mistake. Thorough testing and frequent inspections of radio equipment may detect and fix frequency mismatches before they cause communication problems.

- **Incorrect Settings**: Radios have several options to customize their performance. However, the many choices and customizations might cause operational issues. Reviewing and changing radio setup settings may fix problems. Common problems include accidentally altering settings, misinterpreting parameters, or not configuring the radio for the desired context. Training and user manuals are essential for fixing complicated settings that need technical skills. Operators can quickly recover from configuration problems with frequent inspections and maintenance, and well-documented reset processes. Advanced radio systems also include self-diagnostic functions that automatically fix erroneous settings, reducing downtime and improving dependability.

ADDRESSING BATTERY PACK CONCERNS

Baofeng radios are popular for their affordability and utility in various applications. Like all electronic devices, they have their fair share of common issues. Understanding these problems can help users troubleshoot and maintain their radios effectively. Here's a comprehensive guide to some common issues:

1. **Power and Battery Issues:**
   - **No Power**: Ensure that the battery is charged and properly seated in the radio. The terminals should be clean, with no signs of corrosion or damage.
   - **Short Battery Life**: The expected life of a Baofeng radio battery is around 8-12 hours. If the battery life is significantly shorter, try reducing the screen brightness, turning off unnecessary features, or consider replacing the battery.
   - **Battery Won't Charge**: If the battery isn't charging, check the charger and charging cable for damages. If both are functional, the battery itself might be defective.

2. **Audio and Reception Issues:**
   - **Poor Reception**: Weak signals might be due to physical obstructions, distance from the signal source, or antenna issues. Adjust the antenna, change the location, or consider getting a high-gain antenna for better reception.

- **Distorted Audio**: Distorted or unclear audio may result from volume misadjustment, speaker damage, or poor reception. Adjust the volume, move to a clearer reception area, or consider replacing the speaker.

3. **Programming and Software Issues:**
   - **Programming Difficulties**: Users often face challenges programming the radios for the first time. Make sure you are using the correct programming cable and software compatible with your Baofeng model. Reading the manual or watching online tutorials can also help.
   - **Software Compatibility**: If the programming software exhibits incompatibility with the user's operating system, consider running the software in compatibility mode or alternatively, using another computer system.

4. **Hardware and Accessory Issues:**
   - **Defective Microphone or Speaker**: If the built-in microphone or speaker isn't working correctly, using an external microphone or earpiece can be a temporary solution. For a permanent fix, consider replacing the faulty components.
   - **Malfunctioning Buttons:** Buttons can wear out or malfunction over time. Cleaning around the buttons may help, but if the issue persists, professional repair may be necessary.

5. **Maintenance and Care:**
   - **Regular Cleaning**: To clean the radio body, it is recommended to use a cloth that is either dry or slightly moist. Refrain from using abrasive or caustic substances.
   - **Protective Case Usage**: Consider using a protective case to safeguard the radio from drops, dust, and moisture.
   - **Antenna Care:** Refrain from excessively bending or twisting the antenna. If it becomes damaged, consider getting a replacement.

6. **Firmware and Updates:**
   - **Outdated Firmware**: Make sure your radio is running the latest firmware. If not, consider updating it for enhanced functionality and bug fixes.

7. **Operational Errors:**
   - **Incorrect Frequency Entry**: Double-check the frequency entered to ensure it matches the intended channel.

- **Incorrect Settings**: If the radio isn't functioning as expected, reviewing and resetting the configuration settings can often resolve the issue.

Understanding and identifying common issues with Baofeng radios is essential for effective troubleshooting and maintenance. For problems not covered in this guide, refer to the user manual, online forums, or the manufacturer's customer support for assistance. Consistent care and prompt addressing of issues will extend the life and improve the performance of your Baofeng radio.

Discuss Addressing Battery Pack Concerns:

Addressing concerns related to battery packs in devices like Baofeng radios or other electronic gadgets involves understanding the common problems associated with them and how to troubleshoot or prevent these issues effectively. Below is a detailed discussion on addressing these concerns:

1. **Battery Life Issues:**
   - **Short Battery Life:**
     - **Solutions**: Regularly charge and discharge the battery to maintain its capacity. Reduce the device's power consumption by turning off unnecessary features, decreasing brightness, and using power-saving modes.
     - **Prevention**: Adhere to the manufacturer's guidelines while using the battery. It may be prudent to explore the option of acquiring batteries with larger capacities or employing newer technologies such as Li-ion or Li-Po, as they are known to provide extended lifespans.

2. **Charging Problems:**
   - **Battery Doesn't Charge:**
     - **Solutions**: Inspect the charger, charging cable, and battery terminals for damage. Replace any defective component. Sometimes the battery itself is defective and needs replacement.
     - **Prevention**: Handle charging equipment gently, avoid yanking cables, and use chargers and cables that are compatible with the battery's specifications.

3. **Swelling or Leakage:**
   - **Battery Swelling:**
     - **Solutions**: Replace swollen batteries immediately as they pose a risk of exploding or leaking. Do not attempt to use or puncture a swollen battery.
     - **Prevention**: Avoid exposing batteries to high temperatures, overcharging, or physical damage, as these can cause swelling.

- **Battery Leakage:**
  - **Solutions**: Dispose of leaking batteries properly, taking care not to touch the leaked material. Clean the device's battery compartment thoroughly before inserting a new battery.
  - **Prevention**: Store batteries in a cool, dry place and remove them from devices that won't be used for an extended period.

4. **Battery Replacement:**
  - **Finding Compatible Batteries:**
    - **Solutions**: Always replace batteries with those that match the device's specifications. Using incompatible batteries can damage the device and pose safety risks.
    - **Prevention**: Purchase batteries from reputable manufacturers or authorized dealers to ensure compatibility and quality.

5. **Maintenance and Care:**
  - **Routine Inspection:**
    - Conduct regular visual inspections for signs of damage, swelling, or leakage. Replace batteries that show any of these signs.
  - **Proper Storage:**
    - Store batteries in a cool, dry place away from direct sunlight and heat sources.
  - **Charge Cycles:**
    - It is vital to understand the charge cycle limit of the battery and promptly replace it upon reaching the end of its life cycle.

6. **Disposal:**
  - **Environmentally Safe Disposal:**
    - Never dispose of batteries in regular trash bins as they contain harmful chemicals. Use designated recycling or disposal facilities.

PREVENTIVE MAINTENANCE TIPS:

Preventive maintenance is a crucial aspect of ensuring the longevity and optimal performance of your two-way radios, such as the Baofeng UV5R. Neglecting routine maintenance can lead to costly repairs and reduced operational efficiency. Here are some key preventive maintenance tips to keep your radios in top shape:

- **Regular Inspection**: Begin with a visual inspection of your Baofeng UV5R radios. Look for physical damage, loose connections, or signs of wear. Address any issues immediately to prevent them from worsening.

- **Firmware Updates**: Keep the radio's firmware up to date. Manufacturers often release firmware updates to improve functionality, fix bugs, and enhance security.

- **Cleanliness**: Dust and dirt can accumulate on your radio's exterior and interior components. Regularly clean the radio using a soft, lint-free cloth and, if necessary, a small brush to remove debris from crevices.

- **Check Antennas**: Inspect the antennas for damage or loose connections. A damaged antenna can significantly affect signal quality. Replace it if needed.

- **Battery Maintenance**: Properly maintaining your radio's batteries is crucial, and this leads us to the next subtopic.

DIAGRAM OF BAOFENG UV5R RADIO:

A comprehensive understanding of the constituent elements and operational mechanisms of the Baofeng UV5R radio is required to facilitate efficient problem-solving and upkeep. Presented below is an elaborate schematic representation of the fundamental components comprising the Baofeng UV5R radio:

- **Antenna**: The primary function of the antenna is to facilitate the transmission and reception of signals. Regular inspection is crucial as it is an essential component that warrants careful examination for any potential damage.

- **Display Screen**: The display screen shows important information such as channel settings, battery status, and signal strength.

- **Keypad**: The keypad is used to input various commands, including selecting channels, adjusting volume, and activating features.

- **PTT (Push-to-Talk) Button**: The PTT button allows you to transmit your voice over the radio. Press and hold it when speaking, and release it to listen.

- **Volume Control**: The volume control knob or buttons enable you to adjust the audio output of the radio.

- **Battery Compartment**: This is where the radio's battery is inserted. Proper battery care is essential for long-lasting performance.

- **Speaker and Microphone:** The speaker produces audio output, while the microphone allows you to speak into the radio.

- **Charging Port**: The charging port serves the purpose of replenishing the battery's energy via the process of recharging. It is important to use a charger that is suitable to limit the risk of causing any harm.

- **Accessories Connector**: This is where you can connect various accessories, such as headsets and external microphones.

- **Emergency Alert Button**: Some radios, including the Baofeng UV5R, have an emergency alert button that can send out distress signals in case of emergencies.

Familiarizing yourself with this diagram will help you identify and address issues when troubleshooting and maintaining your Baofeng UV5R radio.

STORE BATTERIES PROPERLY

Properly storing your two-way radio batteries is crucial for maintaining their performance and prolonging their lifespan. Whether you use rechargeable or disposable batteries, following these storage guidelines is essential:

- **Keep Batteries in a Cool, Dry Place**: Store your batteries in a cool, dry environment away from direct sunlight and extreme temperatures. High heat can degrade battery capacity, while excessive moisture can lead to corrosion.

- **Charge Before Storage:** Before storing rechargeable batteries, charge them to 50% of their capacity. This precaution helps reduce the risk of over-discharge, which may harm the battery.

- **Use Original Packaging:** Whenever possible, keep batteries in their original packaging or use designated battery cases to prevent short circuits caused by contact with metal objects.

- **Check and Rotate:** Periodically check your stored batteries and rotate them if you have multiple sets. This practice ensures that all batteries get used and charged regularly, preventing degradation due to inactivity.

- **Avoid Freezing Conditions**: Extreme cold can damage batteries, especially rechargeable ones. Avoid storing batteries in freezing conditions, as it can cause leakage or even rupture.

- **Dispose of Old Batteries Properly:** When batteries reach the end of their life cycle, dispose of them in accordance with local regulations. Many areas have recycling programs for batteries.

By following these suggestions, you can ensure that the batteries of your two-way radios are maintained in optimum condition and are ready to function effectively when required.

CLEAN BATTERIES AND TWO-WAY RADIOS REGULARLY

Regular cleaning is an essential part of maintaining your two-way radios, particularly the batteries. Clean radios and batteries not only look better but also function more efficiently and have a longer lifespan. Here are some tips for cleaning these critical components:

Cleaning Radios:

- **Exterior Cleaning**: Wipe the radio's exterior with a gentle, lint-free cloth or microfiber towel. If needed, dampen the cloth with water or a mild cleaner. Avoid moisture in radio ports and apertures.
- **Cleaning Ports**: Use a can of compressed air to blow out dust and debris from ports, such as the charging port and accessory connectors. Be gentle to avoid damage.
- **Screen Cleaning**: If your radio has a display screen, use a screen cleaner or a mixture of distilled water and isopropyl alcohol (70% or higher) to clean it. Apply the solution to a microfiber cloth, not directly to the screen.

Cleaning Batteries:

- **Exterior Cleaning**: Similar to radios, wipe down the exterior of the batteries with a soft cloth. Check for any visible dirt, grime, or corrosion on the battery contacts.
- **Battery Contacts**: When faced with corrosion on the battery contacts, it is advisable to use a small brush, such as a soft toothbrush, in conjunction with a solution of baking soda and water to gently clean the contacts. Before proceeding with this process, ensure the batteries are removed from the radio.
- **Storing Clean Batteries**: Once your batteries are clean and dry, store them in a clean, dry place following the guidelines mentioned in the previous subtopic.

Regular cleaning not only keeps your radios and batteries looking their best but also ensures optimal performance and helps prevent issues caused by dust, dirt, or corrosion.

Incorporating these maintenance practices into your routine can help keep your two-way radios, particularly the Baofeng UV5R, in top-notch condition. Whether it's preventive maintenance, understanding your radio's components, proper battery storage, or regular cleaning, taking these steps will ensure that your radios are ready for reliable communication whenever you need them..

CHAPTER 9: NAVIGATING THE LEGAL LANDSCAPE

Users should be aware of legal restrictions and considerations when operating these radios within their respective countries. Regulations vary greatly from country to country, and even within regions, so it's essential to understand the specific legal landscape you are navigating. Engaging with the amateur radio community, taking advantage of educational resources, and consulting with regulatory bodies are proactive steps users can take to ensure they operate within the confines of the law. Below are general areas to consider:

Licensing

- Amateur Radio License: Many Baofeng radios operate on frequencies reserved for amateur radio (ham radio). In many jurisdictions, an amateur radio license is required to transmit on these frequencies legally. Obtaining a license typically involves passing an examination that tests your understanding of radio theory, operation, and regulations.

- Commercial License: If you use Baofeng radios for commercial purposes, you might need a different type of license.

Frequency Bands

- Illegal Frequencies: Baofeng radios can transmit on a broad spectrum of frequencies. Some of these frequencies are illegal for public use, reserved for military, aviation, or emergency services. Transmitting on unauthorized frequencies can result in fines, imprisonment, or confiscation of equipment.

- Programmable Frequencies: Since the radios are programmable, users must ensure they are only operating on frequencies they are licensed or allowed to use.

FCC Regulations (For Users in the U.S.)

- In the United States, the Federal Communications Commission (FCC) regulates radio communications. Baofeng radios have been subject to scrutiny, and some models have been banned due to their ability to transmit on unauthorized frequencies.

- Users need to ensure that their radios are FCC compliant. Non-compliant radios are illegal to import, sell, or operate.

International Regulations

- Different countries have their regulatory bodies with specific sets of rules and requirements for radio use. Always adhere to your country's regulations regarding radio transmissions, licensing, and equipment certification.

Legal Responsibilities

- Operators are responsible for their transmissions. This includes avoiding interference with other services and ensuring that communication is carried out ethically and legally.
- Violations of laws and regulations can result in severe penalties, including fines and imprisonment.

Responsible Use

- Familiarize yourself with the specific laws and regulations applicable to radio use in your area.
- Program your radio to operate only within legal bands and frequencies.
- Obtain the necessary licenses before transmitting.

UNDERSTANDING FCC REGULATIONS

The Federal Communications Commission (FCC) is a U.S. government agency responsible for regulating communications by radio, television, wire, satellite, and cable across the United States. Here are key elements to understand about FCC regulations:

1. Scope and Jurisdiction:

- The FCC oversees the communication sector in the U.S., ensuring it runs smoothly, remains accessible, and operates legally.
- It has jurisdiction over all 50 states, the District of Columbia, and U.S. possessions.

2. FCC's Primary Goals:

The Federal Communications Commission (FCC), an essential regulatory body within the United States government, functions with a core set of objectives crucial to its purpose and overarching responsibilities in the field of communications and technology. These objectives demonstrate the FCC's dedication to defining and supervising the continuously evolving communication environment of the country, with the ultimate aim of enhancing societal well-being.

- **Promoting Competition**: The FCC seeks to promote competition in communications services, understanding that competition drives innovation, cost-efficiency, and service quality improvement. It promotes competition to provide customers a variety of options and encourage service providers to improve pricing, performance, and accessibility, thereby benefiting consumers and industry growth.

- **Encouraging Innovation:** In today's rapidly changing technological world, innovation drives development. The FCC endeavors to promote new technology and services, keeping the communications industry flexible and adaptive to customer requirements and preferences.

- **Protecting Consumers**: Protecting consumers from harmful practices is a significant aim of the FCC. This involves adopting and implementing laws to protect customers' privacy and security, and providing accurate and clear service information, thus enabling consumers to make informed choices in a secure and trustworthy digital environment.

- **Public Safety and Homeland Security:** Public safety through dependable communication services is a top priority for the FCC, especially during catastrophes and homeland security crises. In this area, the Commission helps maintain robust and responsive communication networks for first responders, crisis management, and national security.

3. **Radio Frequency (RF) Devices:**
   - Devices that produce radio frequency, such as cellphones, Wi-Fi routers, and two-way radios, are required to adhere to precise technical criteria to reduce the occurrence of detrimental interference, ensure proper operation and safety of devices, and safeguard individuals from excessive radio frequency (RF) exposure.

4. **Licensing**:
   - The FCC issues licenses for various services, including amateur radio, commercial radio and TV broadcasting, and satellite transmissions. Individuals or companies wishing to engage in activities within these designated regions are required to submit an application and obtain a license, entailing certain duties and responsibilities.

5. **Enforcement:**
   - The FCC enforces regulations, handling issues ranging from unlicensed broadcasts to indecent programming. Violators can face substantial fines, equipment confiscation, and, in severe cases, imprisonment.

6. **Spectrum Allocation:**
   - The FCC manages and allocates the electromagnetic spectrum, ensuring that different types of communication services have the necessary bandwidth to operate effectively without interfering with each other.

7. **Public Input and Participation:**
   - The FCC allows and encourages public participation in its decision-making processes, often seeking comments from citizens, industry stakeholders, and other parties on proposed rule changes or other initiatives.

How to Stay Compliant:

In an increasingly connected world where technology dominates our everyday lives, compliance with rules is crucial, especially when using FCC-regulated equipment. This extended guide to FCC compliance can help you navigate this complex terrain and comply with regulations:

- **Thorough Understanding of Regulations:** The first step in maintaining compliance is to develop a deep understanding of the FCC's regulations and guidelines. Access the FCC's official website or relevant documentation to stay informed about specific requirements pertaining to your devices or operations.
- **Certified Devices**: FCC-certified devices are essential for compliance. The certification process entails extensive testing to ensure the gadget meets FCC technical criteria.
- **Acquire Necessary Licenses**: Obtain a license from the FCC for certain operations, such as broadcasting, spectrum usage, or operating in specific frequency bands. Research the licensing requirements relevant to your operations, submit the necessary applications, and await the issuance of the license.
- **Adherence to Operational Guidelines**: Ensure that your equipment operates within the FCC's specified technical and operational parameters to avoid harmful interference with other users and non-compliance with regulations. Familiarize yourself with the operational guidelines pertaining to your specific devices, and regularly audit your operations to ensure ongoing adherence.

In conclusion, staying compliant with FCC regulations is a multifaceted process that demands diligence and a thorough understanding of the rules and guidelines governing electronic communications. By comprehensively educating yourself on these regulations, verifying the certification of your devices, obtaining the required licenses, and strictly adhering to operational parameters, you can ensure that your use of FCC-regulated technology is both legally sound and conducted in a manner that respects the rights and operational needs of others in the spectrum.

LICENSING DEMYSTIFIED: A STEP-BY-STEP GUIDE

Understanding licensing can indeed be complex due to the many different types of licenses and the various rules that apply to each. While licensing requirements can vary from industry to industry, below is a general step-by-step guide to demystify the licensing process, with a focus on communication and radio licensing:

Step 1: Determine the Type of License Needed

- **Understand the Purpose**: Identify the specific use case for your radio communications (e.g., amateur, commercial, broadcasting, public safety).

- **Check Your Device**: Certain devices may only operate on specific bands or frequencies and may require different licenses.

Step 2: Education and Preparation

- **Study Material**: Obtain and study relevant material for the license exam if applicable (mainly for amateur radio licenses).

- **Training Courses**: Enroll in training courses or attend workshops/seminars if available and necessary.

Step 3: Application Process

- **Fill Out Application**: Ensure that you diligently fill out the necessary application forms, ensuring that the information provided is both precise and comprehensive. This includes furnishing correct details about your personal background as well as giving a thorough description of the intended purpose for which the radio equipment will be used.

- **Prepare Supporting Documents**: Gather and submit any necessary supporting documents (e.g., identification, proof of eligibility).

- **Pay Fees**: Pay any associated application or examination fees.

Step 4: Examination (For Amateur Radio Licenses)

- **Take Exam**: Sit for the license examination if required (e.g., Technician, General, or Extra Class license exams for amateur radio in the U.S.).

- **Pass Exam**: Achieve the necessary passing score on the examination to qualify for the license.

Step 5: Await Approval

- **Processing Time**: Wait for the licensing authority to process your application, which can vary in duration.

- **Status Updates**: Check the status of your application online or through other provided channels.

Step 6: License Issuance

- **Receive License**: Once approved, you'll receive your license, often in digital or physical form.
- **Understand Terms**: Review and understand the terms, conditions, and responsibilities associated with your license.

Step 7: License Renewal and Maintenance

- **Renewal**: Be aware of the expiration date of your license and the process for renewal.
- **Updates**: Notify the licensing authority of any changes to your information or status.
- **Compliance**: Ensure ongoing compliance with the rules and regulations associated with your license.

Specific Considerations:

- **Licensing Body**: Licensing procedures might differ depending on the governing body or agency overseeing the licensing in your jurisdiction (like FCC in the U.S.).
- **International Usage**: If you plan to use radio equipment internationally, be aware of reciprocal licensing agreements or the need for additional licenses in other countries.

ETHICAL CONSIDERATIONS FOR RADIO USERS

Radio users, whether they are amateur operators, professional broadcasters, or commercial entities, must adhere to a set of ethical considerations to ensure respectful, lawful, and safe communication practices. Below are key ethical considerations that radio users should take into account:

1. **Respect for Laws and Regulations:**
   - **Adherence**: Legal compliance is the cornerstone of appropriate radio communication. Rules regarding radio frequency in different nations and regions prevent electromagnetic spectrum interference and disorder. These regulations may encompass frequency, power, equipment, and usage restrictions. As responsible radio users, understanding and following these local laws is imperative. By utilizing the radio spectrum efficiently and orderly, you minimize interference and disputes with other users.
   - **Licensing**: Respecting radio communication rules also entails obtaining and maintaining licenses and authorizations. Many nations mandate radio operators to have licenses to operate on specified frequencies or use certain equipment. These licenses often come with stipulated requirements. Severe infractions such as failing to acquire permits or utilizing unlawful frequencies may lead to penalties, equipment seizure, or even incarceration.

2. **Responsible Communication:**

- **Avoid Harmful Content**: Refrain from transmitting harmful, threatening, or harassing content. This includes hate speech, misinformation, and content that might incite violence or discrimination.

- **Respect Privacy**: Avoid intruding on others' privacy. Do not eavesdrop or intercept communications that are not intended for you.

- **Avoid Interference**: Take care not to interfere with other legitimate communications, especially emergency transmissions.

3. **Respect for the Community:**

- **Courtesy**: Practice polite and courteous communication. Respect the norms and etiquette established within the radio community.

- **Support**: Assist newcomers and be cooperative with fellow radio operators, providing help and guidance where necessary.

4. **Emergency Situations:**

- **Priority for Emergency Communications**: Always give priority to emergency transmissions. If you hear a distress call, offer assistance or clear the frequency for emergency use.

- **Volunteer**: Consider participating in emergency communication networks and volunteering your skills and equipment for public service during crises.

5. **Technical Proficiency:**

- **Education**: Continuously educate yourself on the technical aspects of radio operation and stay updated on best practices and emerging technologies.

- **Maintenance**: Ensure your equipment is well-maintained and operates efficiently to prevent unintentional interference or poor communication quality.

6. **Integrity and Honesty:**

- **Transparency**: Clearly identify yourself using the proper call signs and be honest about your intentions and purposes while communicating.

- **No Deceptive Practices**: Avoid engaging in deceptive practices or misrepresentation while on the air.

7. **Consideration for Content:**

- **No Obscenity or Profanity**: Refrain from using obscene language or discussing inappropriate topics, keeping in mind that radio communications can be heard by various audiences, including children.

- **Avoid Commercial Content**: For amateur radio operators, avoid engaging in communications for commercial or financial gain.

8. **International Goodwill:**

- **Foster Friendship**: Use radio communications to promote international goodwill and understanding, respecting the cultural and legal differences of operators from around the world.

- **Abide by International Regulations**: When communicating internationally, adhere to the regulations and standards established for international radio communication.

CHAPTER 10: BUILDING COMPREHENSIVE COMMUNICATION PLANS

Baofeng Radio is a popular brand of affordable two-way radios, commonly used in a variety of settings, including outdoor recreational activities, security details, and emergency services. When developing a comprehensive communication plan for the use of Baofeng Radios, there are several elements that should be carefully considered to ensure effective and efficient communication. Below, I've outlined a generic approach for building communication plans for devices like Baofeng radios.

Communication Plan Components:

1. **Communication Objectives:**
   - Clearly define the purposes of communication: emergency coordination, routine updates, operational control, etc.
   - Specify expected outcomes and goals for radio communication.

2. **Target Audience:**
   - Identify who will be using the radios.
   - Consider the users' level of technical proficiency and training needs.
   - Adjust communication protocol based on audience characteristics.

3. **Equipment Selection and Distribution:**
   - The task at hand involves selecting suitable Baofeng Radio models, considering their features, range capabilities, and battery life.
   - Assign radios to individuals or teams, considering their roles and responsibilities.
   - Keep an inventory of all equipment and accessories.

4. **Training and Skill Development:**
   - Conduct training sessions for users on operating the radios and understanding the communication protocols.
   - Offer refresher courses and ongoing support to ensure users stay proficient.

5. **Channel Planning:**

- Determine the number of channels required for various communication needs.

- Assign and label channels for specific purposes: emergency, general communication, coordination, etc.

- Establish a system for monitoring and maintaining channels to prevent interference and unauthorized access.

6. **Communication Protocols:**

- The objective is to establish uniform protocols for the initiation and response to radio communication.

- Implement codes or shorthand to convey information efficiently and clearly.

- Designate "communication leaders" responsible for directing and overseeing radio communication during events or operations.

7. **Emergency Response Procedures:**

- Establish procedures for using radios in emergency situations.

- Develop a clear and concise protocol for reporting incidents, requesting assistance, and coordinating response efforts.

8. **Maintenance and Upkeep:**

- Plan regular equipment inspections and maintenance to ensure radios remain in good working condition.

- Establish a process for troubleshooting and repairing faulty equipment.

- Update software and firmware as needed

9. **Feedback and Improvement:**

- Collect feedback from users on the effectiveness and efficiency of the communication plan.

- The objective is to conduct an analysis of communication patterns observed during events or activities to identify potential areas for improvement.

- Regularly update and enhance the communication strategy to effectively handle challenges and accommodate evolving demands.

Implementation:

In any organization or setting where reliable communication is pivotal, the successful implementation of a well-structured communication plan is essential. To ensure that this plan not only exists on paper but thrives in practice, a strategic approach must be adopted, involving a series of integral steps that go beyond mere planning.

Here, we delve into an extended discussion of these key implementation phases: Training Sessions, Dry Runs, and the ongoing Review and Revision process.

1. Training Sessions: Building Proficiency and Confidence

Training sessions are a cornerstone of effective communication implementation. These workshops provide an invaluable platform for designated users to enhance their skills, gain confidence, and understand the nuances of the communication equipment and protocols. It's during these sessions that participants acquire a deeper understanding of the tools at their disposal, enabling them to make the most of their communication resources.

Training sessions may encompass a variety of subjects, ranging from fundamental equipment functioning to more sophisticated communication tactics. The objective is to equip users with the necessary information and practical abilities to effectively handle various situations. Training should be seen as a continuous and dynamic process rather than a singular occurrence, as it is essential for users to consistently enhance their skills and stay informed about the newest advancements in communication technology.

2. Dry Runs: Realistic Simulation for Preparedness

To further solidify the effectiveness of the communication plan, the next step involves performing dry runs or practice drills. These drills serve to familiarize users with the equipment and protocols in a controlled, simulated environment. During these exercises, users are placed in scenarios that replicate real-world situations, enabling them to practice their communication skills and evaluate their ability to respond to various challenges.

Dry runs are essential for stress-testing the communication plan. They allow users to identify potential weaknesses and areas for improvement while building confidence in their ability to handle unexpected situations. In essence, dry runs serve as a bridge between theory and real-world application, offering a risk-free environment in which to fine-tune communication techniques.

3. Review and Revise: The Continuous Improvement Cycle

Communication plans, once established, are not static documents. They require continuous evaluation and refinement. The review and revision process is an ongoing commitment to assessing the effectiveness of communication strategies and making necessary adjustments. Regularly scheduled evaluations help identify emerging communication needs, technological advancements, or changes in the operating environment.

The information gathered from real-world experiences, training sessions, and dry runs is invaluable in this process. It informs decisions about protocol changes, equipment upgrades, and the introduction of new communication tools. Moreover, it ensures that the communication plan remains agile and adaptable, capable of addressing evolving challenges and opportunities.

The effective implementation of a communication plan is not a one-off endeavor but a dynamic process that involves training to build proficiency, dry runs to enhance preparedness, and a continuous cycle of review and revision. This approach ensures that the organization or team is well-prepared to handle a wide range of communication scenarios, adapt to changing circumstances, and maintain a robust and reliable communication network.

Continuous Improvement:

It is important to consistently evaluate and revise the communication strategy to effectively accommodate technology improvements, user input, and evolving operational needs. Implement continuous training and support programs to actively involve users, thereby enhancing their proficiency and comfort levels with the radios and communication protocols.

DESIGNING EFFECTIVE COMMUNICATION STRATEGIES

Creating effective communication strategies for Baofeng radios is crucial in ensuring that users can communicate clearly and efficiently in various situations. An effective communication strategy is based on careful planning, understanding the users' needs, and adapting to the operational environment.

Below are essential steps for designing effective communication strategies:

1. **Understand the Users' Needs:**
   - Identify the users' requirements, expectations, and challenges concerning communication.
   - Understand their level of expertise, the contexts in which they'll use the radios, and their communication preferences.

2. **Select Suitable Devices:**
   - Choose Baofeng radio models that meet users' needs, considering factors like range, battery life, and durability.
   - Invest in necessary accessories like earpieces, microphones, and antennas to enhance communication.

3. **Develop User-Friendly Protocols:**

   - Establish clear, straightforward communication protocols, reducing complexity and confusion among users.

   - Implement codes, call signs, and standard phrases to facilitate quick and precise communication.

4. **Allocate Channels Wisely:**

   - Assign channels for different purposes, like general communication, emergency calls, and coordination among specific groups.

   - Establish guidelines for channel use and switching, minimizing interference and confusion.

5. **Provide Training and Support:**

   - Conduct training sessions to familiarize users with the equipment and protocols.

   - Continuous assistance and feedback methods should be provided to swiftly address any queries and concerns raised by users.

6. **Test and Refine Communication Strategies:**

   - Conduct periodic assessments of the efficacy and efficiency of the communication system across various scenarios and settings.

   - Acquire feedback from users and analyze communication records to discover potential areas for enhancement and adaptation.

7. **Ensure Security and Privacy:**

   - Implement security measures to protect communication from unauthorized interception and use.

   - Educate users on maintaining privacy and confidentiality while using the radios.

8. **Plan for Emergencies and Contingencies:**

   - Develop specific protocols for emergency communication, ensuring rapid response and coordination.

   - Prepare for possible communication challenges, like signal loss or equipment failure, and train users on handling these situations.

9. **Maintain and Upgrade Equipment Regularly:**

   - Implement a maintenance schedule to keep devices in optimal condition.

- Stay knowledgeable about the most recent advancements and enhancements available for the radios, and include them when deemed appropriate.

Implementation Steps:

- **Develop Communication Guidelines:**

Write detailed communication guidelines, including all established protocols, codes, and channel allocations.

Disseminate the recommendations to all users, accompanied by appropriate explanations and demonstrations as required.

- **Conduct Training Programs:**

Organize regular training sessions for new and existing users, covering equipment use, communication protocols, and troubleshooting techniques.

- **Monitor and Assess Communication:**

Assign individuals or teams to oversee communication, ensuring adherence to protocols and identifying areas for improvement.

Continuous Improvement:

- Regularly revisit and revise the communication strategies, incorporating user feedback, technological advances, and lessons learned from practical use.
- The aim is to actively involve users in continuous dialogues on communication obstacles and remedies, hence promoting a cooperative approach towards enhancing the efficacy of communication.

UTILIZING BAOFENG RADIOS IN PLANS

Utilizing Baofeng radios effectively in various planning contexts requires an understanding of how the radios function, who will be using them, and for what purpose. The application of these radios can be versatile, covering emergency management, event planning, field operations, and more. Here's a general guide on how to utilize Baofeng radios in different plans:

Emergency Management Plan:

1. **Preparation:**

   - Select radios with sufficient range and features like emergency alerts.

   - Establish dedicated channels for emergency communication.

2. **Training:**

   - Train team members on emergency protocols, channel use, and the use of codes.

3. **Emergency Protocols:**

   - Develop swift communication procedures for different types of emergencies, such as natural disasters, medical emergencies, or security issues.

Event Planning:

1. **Coordination:**

   - Use radios for real-time coordination among event organizers, security, and staff.

   - To minimize the occurrence of crosstalk and confusion, it is advisable to assign distinct channels to various departments or teams.

2. **Crowd Management:**

   - Equip security and volunteer staff with radios to manage crowds effectively and respond promptly to issues.

3. **Setup and Teardown:**

   - Enhance communication between the teams involved in the construction and dismantling of event infrastructure.

Field Operations:

1. **Team Communication:**

   - Enable field teams to stay in contact with each other and with base operations.

   - Use radios for coordination, reporting findings, and requesting support.

2. **Safety Protocols:**

   - Develop procedures for reporting safety incidents and requesting emergency assistance.

Security Detail:

1. **Patrolling:**

   - Equip security personnel with radios to facilitate communication while patrolling.

   - Generate codes to categorize various sorts of security issues.

2. **Access Control:**

   - Use radios to authenticate access rights and establish communication with security checkpoints.

Incorporating Baofeng Radios into Plans:

1. **Needs Assessment:**

   - Evaluate the communication requirements and constraints of your plan.

   - Select appropriate Baofeng radio models and accessories.

2. **Channel Allocation:**

   - Allocate and label channels for specific uses and teams.

   - Establish guidelines for channel use to prevent misuse and interference.

3. **Training and Familiarization:**

   - Conduct training sessions to familiarize users with radio operations and communication protocols.

   - Offer hands-on practice opportunities and provide support materials for reference.

4. **Protocol Development:**

   - Establish clear communication protocols, including the use of codes and call signs.

   - Define procedures for regular communication, emergency calls, and situation reports.

5. **Maintenance and Support:**

   - Set up a system for maintaining, charging, and storing radios.

- Provide technical support to assist users with radio-related issues during operations.

Implementation:

1. **Test and Drill:**

   - Conduct regular tests and drills to practice using radios in various situations defined in the plans.

   - Identify and resolve issues or gaps in communication that may arise during these drills.

2. **Feedback Loop:**

   - Create mechanisms for collecting feedback from radio users regarding device performance, ease of use, and communication effectiveness.

   - Implement improvements based on feedback and lessons learned from actual operations.

Continuous Improvement:

- Review the use of radios post-implementation, analyzing their effectiveness in facilitating communication during planned operations.

- Adjust protocols, training programs, and device configurations based on the insights gained from actual use.

CASE STUDIES AND EXPERT INSIGHTS

Case studies and expert insights on Baofeng radios are instrumental in understanding the diverse applications, user experiences, challenges, and solutions associated with these devices. However, I don't have specific real-time or individual case studies, but I can generate hypothetical ones based on common use cases and expert insights.

Hypothetical Case Studies:

1. **Case Study: Emergency Response During Natural Disasters**

   - **Scenario**: During natural disaster relief activities, a community-based group effectively employs Baofeng radios to provide seamless communication among the team members.

   - **Insights**:

   - Radios are crucial in areas with network outages.

   - The gadgets serve to enhance coordination efforts in the distribution of relief items, conducting rescue operations, and providing medical help.

2. **Case Study: Security Team Communication at Large Events**

- **Scenario**: Security personnel at a significant cultural event use Baofeng radios to monitor and manage crowd activities and coordinate responses to incidents.
- **Insights**:
  - Radios allow instant communication among team members.
  - Quick response to emerging security issues is enhanced.

3. **Case Study: Outdoor Adventure and Safety Communication**
   - **Scenario**: An adventure club uses radios while trekking in remote areas, maintaining group cohesion and ensuring safety.
   - **Insights**:
     - In locations without cellular reception, radios are lifesavers.
     - Clear communication protocols enhance group safety and coordination.

Expert Insights:

1. **Affordability & Accessibility:**

Insight: Experts appreciate Baofeng radios for being cost-effective, making them accessible to a broader user base. They are ideal for users who need reliable communication tools without breaking the bank.

2. **Ease of Use:**

Insight: With user-friendly interfaces and straightforward functions, these radios are easy to operate, even for individuals without technical expertise. Experts often recommend them for beginners and casual users due to their simplicity.

3. **Limitations & Legal Constraints:**

Insight: Experts caution users to be aware of the limitations, including potential interference with other communication devices. Users should also consider legal restrictions and licensing requirements in their respective jurisdictions.

4. **Training & Protocol Development:**

Insight: Developing clear communication protocols and investing in user training are essential for maximizing the effectiveness of Baofeng radios. Experts recommend regular practice and drills to familiarize users with the equipment and procedures.

Considerations:

In a world marked by constant connectivity and instant communication, Baofeng radios have carved out a niche as versatile tools for both personal and professional use. Whether it's for outdoor adventures, emergency preparedness, or maintaining seamless communication within a business, understanding the context, technical expertise, and legal compliance are pivotal considerations to ensure these radios serve their purpose effectively.

1. **Understanding the Context**: The successful utilization of Baofeng radios begins with a profound understanding of the context in which they will be employed. Whether you're planning to use these radios for recreational activities like hiking or camping, or in more demanding situations such as search and rescue operations, comprehending the specific environment and user communication needs is paramount. Each setting presents unique challenges, be it the topography, range of communication required, or potential obstacles like buildings or natural terrain. A nuanced understanding of these factors is key to choosing the right model and using it optimally.

2. **Technical Expertise**: While Baofeng radios are generally designed to be user-friendly, providing adequate training is essential to ensure users can harness their capabilities effectively. Technical expertise ensures users can adapt to different scenarios, make the most of features like dual-band frequencies, and troubleshoot common issues. Training also familiarizes users with the radio's interface, including programming, channel selection, and using advanced features like repeaters. For more specialized applications, such as amateur radio operators or emergency responders, additional training may be necessary to unlock the radios' full potential. By investing in user education, individuals and organizations can harness the power of Baofeng radios with confidence and efficiency.

3. **Legal Compliance**: Regional laws and regulations govern radio communication, including Baofeng radios. Users must strictly follow these regulations for safe and legal use. Radio communication legality may depend on licensing, frequency allotment, and power output. Not following these rules might result in penalties, equipment confiscation, or criminal prosecution. To prevent legal issues, one must grasp local, national, and international radio communication legislation.

In conclusion, Baofeng radios offer a valuable means of communication across various contexts, but their effective use hinges on understanding the context of application, acquiring technical expertise, and ensuring legal compliance. By carefully considering these facets, users can maximize the utility of these radios, foster effective communication, and avoid potential pitfalls. Whether for personal adventures or professional communication needs, Baofeng radios, when used responsibly and skillfully, become indispensable tools in the modern communication toolkit.

CHAPTER 11: EFFICIENT COMMUNICATION WITH TRIGRAMS

"Trigrams" in the context of communication could be interpreted in two different ways: the term might refer to a sequence of three letters, words, or symbols, often used in natural language processing and cryptography, or to the eight three-line combinations of broken and unbroken lines (known as "gua") found in the I Ching, an ancient Chinese text that is often used for divination purposes.

1. Efficient Communication with Textual Trigrams:

When discussing trigrams in terms of natural language processing or cryptography:

Key Features:

- **Predictive Text and Auto-correction**: Trigrams assist in creating predictive text models by analyzing the probability of the third word given the previous two. This process enhances efficient communication by suggesting relevant words to users while typing.
- **Search Optimization**: Search engines can utilize trigram models to understand and match queries with relevant content effectively, thereby fetching more accurate search results.

Best Practices:

- **Clean and Preprocess Data**: To achieve optimal performance in trigram communication models, it is essential to ensure that the text data used is of high quality, devoid of any extraneous noise, and subjected to proper preprocessing techniques.
- **Use of Smoothing Techniques**: Incorporate smoothing techniques to handle trigrams that don't appear in the training data but might appear in real-life communication scenarios.

2. Efficient Communication with I Ching Trigrams:

When discussing trigrams as found in the I Ching:

Key Features:

- **Symbolic Representation**: Each trigram has a distinct semantic significance and symbolizes a range of natural elements and ideas. Comprehending these symbols has the potential to enhance the conveyance of intricate concepts using uncomplicated symbols.

- **Mindfulness and Reflection**: Engaging with these trigrams encourages mindfulness and reflective thinking, fostering a different kind of communication that is more introspective and thoughtful.

Best Practices:

- **Study and Understand Symbols**: To communicate effectively using I Ching trigrams, one should invest time in studying and understanding the meaning and interpretation of each trigram.

- **Practice Mindful Communication**: Engage in introspective and conscious communication methodologies, using trigrams as a mechanism for deep reflection and profound understanding, rather than relying solely on conventional linguistic expressions.

INTRODUCTION TO TRIGRAMS IN RADIO COMMUNICATION

Trigrams in radio communication may not refer to a well-known or established term, and it could be possible that you might be referring to something specific in a niche area, or perhaps you might have misinterpreted the term.

In communications and information processing, "bigrams," "trigrams," etc., often refer to sequences of two, three, or more characters or bits, and they are used in various applications, including natural language processing, cryptography, and coding theory.

Below, I'll discuss a general idea of how sequences like trigrams might be related to radio communications:

1. Signal Modulation:

Trigram Sequences: If trigrams are sequences of three bits (000, 001, 010, etc.), they could be used in digital radio communications to represent different modulation states or symbols. In digital modulation schemes, bits are represented by changes in amplitude, frequency, or phase of the carrier wave. A trigram could represent a specific change or set of changes in the carrier wave to transmit three bits of information at once.

2. Coding and Error Correction:

Error-Correction Codes: Trigrams or bit triplets could be employed in error detection and correction schemes. For example, in a simple repetition code, each bit is transmitted three times (a bit '0' would be transmitted as

'000', and a bit '1' would be transmitted as '111'). The receiver can then use majority voting to correct errors that occur during transmission.

3. Protocol Design:

Control Sequences: Sequences of bits, including trigrams, could be used as control characters or sequences in communication protocols to signal the start or end of a message, error conditions, or other events in a communication system.

4. Cryptography:

Cipher Systems: In secure radio communication, sequences of bits like trigrams might be used as elements in cryptographic algorithms to encrypt and decrypt messages.

5. Application in Data Compression:

Data Representation: Trigrams can be used to efficiently represent and compress data that is being transmitted over radio waves.

COMPLETE TRIGRAM LIST FOR PRACTICAL USE

When referring to trigrams in a practical use scenario, one might commonly think about their application in language modeling, cryptography, or data compression, among other fields. Trigrams can refer to sequences of three characters, symbols, or bits, and their interpretation can vary significantly based on the context in which they are used.

Below is a general approach to understanding the different applications of trigrams:

1. Language Modeling:

- **Trigram Model**: In Natural Language Processing (NLP), trigrams are sequences of three adjacent words or characters. The trigram model predicts the likelihood of a word based on the two preceding words. This model is widely used for speech recognition, text generation, and other applications.
- **Practical Use**: When building a trigram model, you would typically extract all unique trigrams from your dataset and count their occurrences. This dataset-driven approach ensures that the list of trigrams is practically useful for the specific application.

2. **Cryptography:**

- **Cipher Systems**: In cryptography, trigrams might be used in the context of creating keys or in cipher systems that encrypt and decrypt messages.

- **Practical Use**: A complete list of trigrams (assuming binary trigrams) would consist of all $2^3 = 8$ possible combinations of three bits: 000, 001, 010, 011, 100, 101, 110, and 111.

3. **Data Compression:**

- **Encoding**: Trigrams can be used to represent and compress data efficiently. For example, Huffman coding might use trigrams to represent frequent sequences of data with shorter codes.

- **Practical Use**: In this case, the complete list of trigrams would depend on the data being compressed. You would identify frequent sequences of three characters or bits and assign short codes to them.

4. **Bioinformatics:**

- **Genetic Sequences**: In bioinformatics, trigrams (or 'codons' in this context) are sequences of three nucleotides, which code for specific amino acids in protein synthesis.

- **Practical Use**: There are 64 possible codons or trigrams in the genetic code, each representing a specific amino acid or a stop signal for protein synthesis.

5. **Signal Processing & Communication:**

- **Symbol Representation**: Trigrams might be used as symbols in communication protocols or modulation schemes.

- **Practical Use**: The list of trigrams would depend on the protocol or scheme. In digital communication, trigrams could be sequences of three bits, with eight possible combinations, as mentioned above.

How to Compile a Trigram List:

The generation of a complete trigram list is an essential undertaking in several academic disciplines, such as natural language processing, encryption, and data analysis. This procedure encompasses many pivotal phases, each of which contributes to the creation of a valuable asset tailored for a particular use. In the following discussion, we will explore the necessary procedures for constructing a trigram list. This list may include trigrams at the character level, the word level, or be extended to other domains such as genomics.

1. **Data Collection**: The initial step in compiling a trigram list is data collection. This stage entails gathering a substantial dataset that is pertinent to the specific application at hand. The quality and relevance of the dataset are crucial, as they directly impact the utility and accuracy of the trigram list. For instance, in natural

language processing, a corpus of text documents or a collection of specific texts related to the target domain is essential.

2. **Trigram Extraction**: Once a substantial dataset has been procured, the next step is trigram extraction. Trigrams are sequences of three adjacent elements, which could be characters, words, bits, nucleotides, or any other units of interest. This stage involves breaking down the dataset into these three-element sequences. For instance, in text analysis, this would entail creating trigrams by sliding a three-character window over the text, while in genomics, it would involve extracting trigrams from DNA sequences.

3. **Frequency Analysis**: Once a collection of trigrams has been acquired from the dataset, the following crucial step involves conducting frequency analysis. It is vital to comprehend this stage in order to ascertain the prevalence of certain trigrams within the dataset. Through the process of quantifying the occurrence rate of each trigram, valuable insights may be obtained on their importance and pertinence. Frequently occurring trigrams tend to encapsulate significant information or patterns, while seldom occurring trigrams may possess relatively less relevance.

4. **Sorting & Pruning**: To make the trigram list more manageable and focused, the final step involves sorting the trigrams by their frequency and pruning those that are either irrelevant or occur too infrequently to be of practical use. Pruning helps in reducing the complexity of the trigram list, making it easier to work with in subsequent applications. The threshold for pruning can vary depending on the specific goals and requirements of the project.

In conclusion, compiling a trigram list is a systematic process that starts with data collection, proceeds through trigram extraction, frequency analysis, and culminates in the sorting and pruning of trigrams. The resulting trigram list serves as a valuable resource, enabling various applications to analyze patterns, make predictions, or enhance understanding within their respective domains. The quality of the dataset and the efficacy of the pruning process significantly influence the utility of the trigram list in practical applications.

TIPS AND BEST PRACTICES FOR USING TRIGRAMS

General Practices:

- **Understand Context**: Know the domain and context in which you are using trigrams, as practices can vary significantly from language modeling to cryptography.

For Language Modeling and NLP:

- **Large Datasets**: Use sufficiently large and diverse datasets to capture a wide range of trigrams.
- **Handle Unknown Trigrams**: Implement a method for handling unseen trigrams during training, like smoothing or back-off models.
- **Balanced Data:** Ensure your dataset represents various styles, tones, and topics to avoid bias.
- **Data Cleaning**: Remove or handle special characters and numbers appropriately.

For Cryptography:

- **Secure Keys**: If trigrams are part of key generation, use secure and random generation methods.
- **Avoid Patterns**: Patterns in keys can lead to vulnerabilities; ensure randomness.
- **Key Management**: Implement secure key management and storage practices.

For Data Compression:

- **Frequency Analysis**: To achieve optimal compression, it is essential to comprehend the frequency distribution of trigrams within the dataset.
- **Dynamic Adjustment**: Adjust the trigram-based compression algorithm dynamically based on incoming data patterns.

For Bioinformatics:

- **Understand Codons**: Familiarize yourself with genetic codons if working with DNA sequences.
- **Validation**: Validate the detected trigrams against known biological databases for accuracy.

For Signal Processing:

- **Error Detection**: Implement error detection and correction when using trigrams to represent signal states.

- **Synchronization**: It is important to establish a precise synchronization mechanism between the transmitter and receiver in order to achieve correct interpretation of trigrams.

General Data Handling:

- **Data Privacy**: When working with sensitive data, ensure that privacy is maintained and adhere to relevant regulations.

- **Efficient Storage**: Store trigrams efficiently to minimize memory usage, using appropriate data structures.

- **Optimization**: Optimize the trigram generation and processing algorithm for speed and efficiency.

- **Regular Updating**: Periodically update your trigram lists or models to incorporate new data and trends.

- **Testing**: Conduct thorough testing and validation of your trigram-based system to ensure reliability and accuracy.

- **Documentation**: Document the trigram generation, processing, and usage procedures clearly for future reference and maintenance.

CHAPTER 12: DECODING RADIO SIGNALS AND MESSAGES

Decoding radio signals and messages is a process used in communication technology to translate received radio waves into readable or usable information. The type of radio signals, whether they are analog or digital, dictates the specific decoding process employed. Below is a general overview of decoding radio signals and messages.

Analog Signals:

Analog signals, in the realm of radio communication and beyond, are at the heart of many essential processes, shaping the way information is transmitted, received, and processed. These continuous variations in electromagnetic waves have been instrumental in various aspects of technology and communication. Let's dive deeper into the stages of processing analog signals to appreciate their significance and understand their role in contemporary applications.

1. **Reception:** The journey of an analog signal begins with reception. The first step in this process is the antenna, which serves as the initial receiver of these electromagnetic wave fluctuations. The antenna's primary function is to capture the analog signal, whether it's carrying information from a radio station, a television broadcast, or any other source. The antenna is designed to efficiently pick up these waves, converting them into electrical currents that mirror the incoming signal's characteristics.

2. **Amplification:** Once the analog signal is captured, it is often in a weakened state. To ensure the signal is robust enough for further processing and interpretation, amplification becomes necessary. Amplifiers are employed to boost the signal's strength, improving both its amplitude and quality. This step is critical, especially in situations where the signal may have degraded during transmission, such as over long distances or through interference-prone environments.

3. **Demodulation:** Modulating analog signals superimposes the message signal onto a carrier wave to improve transmission. The analog signal must be demodulated to extract the message. AM, FM, or other modulation systems are utilized for transmission, which determines the demodulation process. Demodulation prepares the carrier wave for processing by extracting its basic information.

4. **Conversion:** Analog signals may need to be converted to digital for analysis, interpretation, or storage. This conversion is usually done using ADCs. By sampling and assigning numerical values to analog waveform samples, these devices convert continuous analog signals to discrete digital representations. To store, handle, and transport analog data more effectively in digital systems like computers, this conversion is necessary.

The above-described processing of analog signals serves as the fundamental basis for many technologies, including radio communication, where the signals are adjusted and improved upon continuously to guarantee the integrity and high quality of the information being delivered. For engineers, technicians, and hobbyists alike, understanding the nuances of this analog-to-digital conversion is crucial because it enables them to fully use the potential of analog signals and smoothly incorporate them into our contemporary digital environment.

Digital Signals:

In the era of modern communication and information technology, digital signals are the lifeblood of our interconnected world. These signals, typically represented as discrete values, primarily as 0s and 1s, underpin the transmission and reception of data in a variety of applications, from everyday internet browsing to complex satellite communication systems. Understanding the journey of digital signals, from their reception to decoding, is essential to grasp the remarkable technology that defines our digital age.

1. **Reception:** Digital signals commence their journey at the reception stage, where they are captured by an antenna or receiver. These signals may originate from diverse sources, including wired connections, wireless networks, or even interplanetary transmissions. The antenna acts as the first point of contact, collecting these binary signals and passing them onto the subsequent stages of processing.

2. **Error Checking:** The reliability and integrity of data transmission are paramount in digital communication. To ensure this, digital signals often incorporate error-checking codes. These codes are embedded within the signal and serve as a form of insurance against data corruption during transmission. Various algorithms are deployed for error detection and correction, enabling the system to identify and rectify errors, thus guaranteeing that the received data remains accurate.

3. **Demodulation:** Demodulation is the next critical stage in digital signal processing. Original data is extracted from the carrier wave. Which demodulation method is used depends on the digital modulation type. QAM, PSK, and FSK are used to encode digital data into carrier waves and decode them on the receiving end.

4. **Decoding:** Decoding follows demodulation of the digital stream. This converts binary data into a format computers can use. The data may need to be converted from binary code to text, graphics, or another format that people or machines can understand.

In essence, digital signals are the backbone of our information age, enabling the seamless flow of data across vast distances and diverse mediums. Their journey from reception to decoding involves intricate processes that are underpinned by advanced technology and mathematical algorithms, ensuring that the data we rely on remains accurate, reliable, and accessible in the digital world that surrounds us.

Message Decoding:

The process of decoding messages is a complex and essential aspect of radio communication. The field encompasses a range of methodologies, instruments, and a profound understanding of the protocols and encryption mechanisms used in communication networks. In this discourse, we will delve into the intricacies of message decoding, elucidating the software, protocols, and encryption elements that make it an essential component of radio communication.

- **Decoding Software:** Special software can be used to decode messages encoded with specific protocols or encryption. Some amateur radio operators and hobbyists use software to decode messages transmitted in Morse code, RTTY, or other digital modes.

- **Protocol Understanding:** Various communication systems use different protocols to organize their messages. A comprehensive comprehension of the protocol is crucial for the effective deciphering process.

- **Encryption:** The decoding of encrypted communications necessitates the use of the appropriate cryptographic key and technique. Deciphering the contents of these communications without the requisite key is a considerable challenge, and in some cases, it may be deemed practically unattainable due to the encryption used.

In essence, message decoding is a dynamic field that blends technical prowess with protocol knowledge and a respect for data security. It empowers individuals to unearth the hidden meanings in radio signals, bridging the gap between encoded information and its real-world significance. Whether preserving the art of Morse code or ensuring the security of sensitive communications, message decoding is an essential component of modern radio communication.

Examples of Radio Signal Decoding:

- **Satellite Communication:** Signals sent from satellites are decoded for various applications, including GPS navigation, satellite television, and weather monitoring.

- **Amateur Radio (Ham Radio):** Enthusiasts use various modes to communicate, including voice, Morse code, and digital signals, all of which require decoding.

- **Emergency Services:** Police, fire, and medical services often use encoded or encrypted radio signals for secure communication.

Legal Considerations:

- The act of intercepting, decoding, or using encrypted or proprietary communications without proper authorization is considered unlawful in several legal countries. Engaging in the unauthorized decryption of signals from cellular networks, encrypted police communications, or satellite feeds might result in legal ramifications.

Before attempting to decode radio signals, one should understand the legal constraints and ethical considerations associated with intercepting and decoding radio communications. Always ensure compliance with local, state, and federal laws regarding radio signal interception and decoding.

IMPORTANCE OF DECODING IN RADIO COMMUNICATION

Decoding in radio communication is crucial for various reasons, facilitating the reliable and accurate transmission and reception of information across different mediums. Below are some significant reasons highlighting the importance of decoding in radio communication:

1. **Data Integrity:** Decoding incorporates error detection and correction algorithms, which are crucial for maintaining the integrity of transmitted data. These algorithms help in recognizing and fixing errors that may have occurred during transmission, ensuring the received message is accurate and reliable.

2. **Secure Communication:** Decoding is integral in the implementation of encrypted communications. It allows authorized recipients to decrypt messages, maintaining the confidentiality and security of the transmitted information. This aspect is particularly crucial for military, governmental, and emergency service communications, where sensitive information requires protection from unauthorized interception.

3. **Effective Data Transmission:** Decoding allows the conversion of received signals into a usable format, enabling effective communication between different devices and systems. This conversion is vital for the functionality of various technologies, from mobile phones and internet communication to broadcasting and satellite navigation.

4. **Facilitation of Various Communication Modes:** Different radio communication modes, like AM, FM, and digital signals, require specific decoding techniques. The ability to decode these signals enables the reception of diverse forms of communication, supporting a wide range of applications, including broadcasting, telecommunication, and data transmission.

5. **Accessibility and Usability:** Decoding enables the transformation of raw signals into human-readable or machine-readable formats. For instance, voice messages can be demodulated and decoded into audible

sounds, while data can be translated into texts, images, or other formats that users can easily interact with and understand.

6. **Global Communication:** The transmission of radio signals across national boundaries is a crucial aspect of global communication, and the capacity to decipher these signals has significant importance. The technology enables the transmission of information on a global scale, including international distress signals, and promotes coordination across borders for a variety of reasons, therefore promoting global connectedness and collaboration.

7. **Support for Technological Development:** The practice and improvement of decoding techniques contribute to the advancement of communication technologies. It enables the development of more sophisticated, reliable, and secure communication devices and systems, driving innovation in the field.

8. **Resource Optimization:** Advanced decoding techniques allow for more efficient use of the available bandwidth, facilitating the transmission of more information within limited frequency spectra. This efficiency is crucial as the radio frequency spectrum is a finite resource, and its optimal use is necessary to support the growing number of wireless devices and services.

COMPREHENSIVE DECODE LIST

Baofeng radios are popular among amateur radio enthusiasts for their affordability and functionality. A Comprehensive Decode List would typically refer to the list of frequencies, tones, and codes that the Baofeng radios can decode or demodulate. This includes various signals transmitted over the frequency bands that the radios can operate on, usually in the VHF (Very High Frequency) and UHF (Ultra High Frequency) ranges.

1. Frequency Bands:

- **VHF (136-174 MHz):** Often used for two-way, line-of-sight voice communications. VHF is common among marine and aviation users, as well as emergency services in some areas.

- **UHF (400-520 MHz):** Ideal for use where VHF signals can't penetrate, like in urban areas with tall buildings or in heavily forested regions.

2. CTCSS/DCS Decoding:

- **CTCSS (Continuous Tone-Coded Squelch System):** A sub-audible tone that allows users to communicate within a specific group without hearing traffic directed to different users on the same frequency.

- **DCS (Digital-Coded Squelch):** Similar to CTCSS but uses digital coding. It provides more code combinations, allowing for finer control over group communication.

3. DTMF Decoding:

- **DTMF (Dual-Tone Multi-Frequency):** Often used for controlling other devices over radio waves, such as repeaters or telephones. Baofeng radios can decode these tones to facilitate this kind of remote control.

4. Voice Scramble and Encryption: Baofeng radios may offer voice scrambling features for basic security. While not as secure as advanced encryption techniques, this can provide a minimal level of privacy for communications.

5. Signaling: Certain models of Baofeng radios include the capability to accommodate various signaling techniques, such as selective calling or alert tones. These functionalities aid users in discerning the identity of incoming callers or in activating a repeater's squelch mechanism.

How to Use Decode Features on Baofeng Radios:

1. **Access the Menu:**

   - Press the 'Menu' button on the radio.

2. **Navigate to Desired Function:**

   - Use the arrow or number keys to find the function you wish to adjust (e.g., CTCSS, DCS).

3. **Select the Function:**

   - Press 'Menu' again to select the function.

4. **Adjust Settings:**

   - Use the arrow or number keys to adjust the settings or enter the required codes.

5. **Confirm Settings:**

   - Press 'Menu' or 'Exit' to confirm and save your settings.

Legal and Ethical Considerations:

- Users should ensure they are operating within the law, only transmitting on frequencies they are licensed to use, and respecting privacy and ethical standards when using decoding and scrambling features.

- Always check your local regulations regarding radio use, as improper use of radios and frequencies can result in fines, legal action, or disruption to essential communication services.

CASE STUDIES: REAL-LIFE EXAMPLES OF DECODING MESSAGES

Decoding messages is an essential aspect in various fields, including telecommunications, computer science, cryptography, and military operations. The following are real-life examples and case studies that involve the decoding of messages:

1. World War II Cryptography:

- **Enigma Machine:**

  - The Enigma machine was used by the German military for the purpose of encoding communications, hence posing considerable difficulties for codebreakers from the Allied forces due to its intricate nature.

  - During World War II, British and Polish cryptanalysts, among them the renowned Alan Turing, used various methodologies and devised equipment such as the Bombe to decrypt communications encoded by the Enigma machine. These efforts played a pivotal role in shaping the result of the war.

2. SETI and Decoding Extraterrestrial Signals:

- **The Wow! Signal:**

  - In 1977, a strong narrowband radio signal was received by Ohio State University's Big Ear radio telescope. Though the source and meaning remain unknown, the signal's unique characteristics prompted extensive analysis and decoding attempts.

3. Aviation Communication:

- **Black Box Decoding:**

  - After air accidents, investigators retrieve and analyze Flight Data Recorders (FDR) and Cockpit Voice Recorders (CVR). The data extracted and decoded from these "black boxes" is crucial for understanding events leading to accidents and improving aviation safety.

4. Digital Forensics and Cybersecurity:

- **Decoding Malicious Payloads:**

  - Security analysts often encounter encoded or encrypted malicious payloads in network traffic or during malware analysis. Decoding these elements is vital for understanding the malware's functionality, origin, and potential countermeasures.

5. Telecommunication and Mobile Networks:

- **SMS Decoding:**

  - Text messages transmitted over cellular networks are often encoded for efficient transmission. Network engineers and developers may need to decode these messages during troubleshooting, development, or optimization processes.

6. Internet Communication Protocols:

- **HTTP/HTTPS Decoding:**

  - In network analysis and cybersecurity, professionals decode HTTP and HTTPS traffic to understand communication between clients and servers, identify potential issues, or detect malicious activity.

7. Emergency Services Communication:

- **Decoding Distress Signals:**

  - Emergency beacons and distress signals often transmit encoded information, including the device's identity and sometimes its location. Rescuers decode these signals to facilitate effective and timely response operations.

Legal and Ethical Considerations:

- The act of deciphering encrypted or confidential communications without proper authorization is considered both unlawful and morally questionable in several legal jurisdictions. Respecting privacy, obtaining permission, and adhering to legal requirements are crucial when deciphering signals within any given situation.

CONCLUSION

A reflection on the journey of Baofeng radios could conclude by acknowledging the accessibility and opportunities these devices have provided to individuals worldwide. Through continuous exploration and learning, users can deepen their understanding and skills in radio communication, engage with like-minded enthusiasts, and possibly even contribute to the broader community of Baofeng and radio communication enthusiasts. Whether one is a beginner or an experienced user, the journey with Baofeng radios offers a pathway for ongoing discovery and mastery in the realm of two-way radio communication.

For those interested in Baofeng radios and two-way radio communication in general, the journey indeed continues. Below are aspects that one might consider for continued exploration and learning:

1. Technological Evolution:

- **Updates and Innovations:** Baofeng continuously works on improving the technology and features in their radios. Understanding these updates is crucial for users to make the most out of their devices.

- **Adaptation to New Standards:** As communication standards and technologies evolve, Baofeng radios may adapt to support newer protocols and standards.

2. Community Engagement:

- **User Communities:** There are online forums, social media groups, and clubs dedicated to Baofeng and two-way radio enthusiasts. These communities are valuable resources for learning and sharing knowledge.

- **Events and Gatherings:** Participating in events organized by these communities can be a valuable experience, offering practical knowledge and networking opportunities.

3. Skill Development:

- **Technical Skills:** Learning how to effectively operate and troubleshoot Baofeng radios, understanding radio frequencies, and getting familiar with radio programming are crucial skills for users.

- **Licensing:** For some, obtaining an amateur radio license might be necessary to operate on certain frequencies legally and responsibly.

4. Exploration of Use Cases:

- **Different Environments:** Baofeng radios are used in various environments, from outdoor recreation to professional settings. Understanding how these radios function in different contexts can be enlightening.

- **Experimentation:** Users might experiment with different accessories, antennas, and setups to enhance the performance and capabilities of their radios.

FINAL THOUGHTS AND AN INVITATION TO SHARE FROM MAXWELL CIPHER

We have reached the end of this book together, and I hope it has armed you with knowledge and confidence.

I leave you with one last testimony of how **"The Baofeng Radio Bible"** can make a difference in someone's adventures.

"'The Baofeng Radio Bible' is more than just a manual; it's a journey into the heart of revolutionary communication technology. Maxwell Cipher imparts his extensive knowledge with both passion and precision, making this book essential for anyone looking to enhance their communication skills in critical situations. His step-by-step guides, practical tips, and tricks make complex concepts approachable, offering radio enthusiasts invaluable insights." **– E. Robinson - Emergency Management Specialist**

If you feel equally inspired, don't miss the moment and immediately share your journey by leaving a feedback on Amazon. Your contribution will help make this book an even more solid and reliable reference in the radio communications community.

Go to the ORDERS section of your Amazon account and click on the "Write a review for the product" button, or scan this QR code.

Thanking you for your trust and support, it's been great spending time with you, and I hope you continue to communicate safely and accurately in every situation.

If you wish, let me hear your voice: info@survivalhorizon.com.

With gratitude, *Maxwell Cipher*

GLOSSARY

- **Affirmative**: Yes.

- **Analog Signals**: Continuous variations in electromagnetic waves used in various aspects of technology and communication.

- **Antenna**: Device designed to capture radio signals and convert them into electrical currents.

- **Battery Care**: Measures to prevent battery swelling or leakage and guidelines for battery replacement.

- **CTCSS (Continuous Tone-Coded Squelch System)**: A sub-audible tone allowing users to communicate within a specific group without hearing other users on the same frequency.

- **DCS (Digital-Coded Squelch)**: Similar to CTCSS but uses digital coding for finer control over group communication.

- **Decoding**: The process of translating received radio waves into readable or usable information.

- **Demodulation**: The process of extracting the original data from a modulated carrier wave.

- **Digital Signals**: Signals represented as discrete values, primarily as 0s and 1s, used in modern communication and information technology.

- **DTMF (Dual-Tone Multi-Frequency)**: Used for controlling other devices over radio waves, such as repeaters or telephones.

- **Encryption**: The process of converting information into code to prevent unauthorized access.

- **Error Checking**: Ensures the reliability and integrity of data transmission by detecting and correcting errors.

- **Licensing**: The process of acquiring the necessary permissions to operate on certain frequencies.

- **Maintenance**: Regular care and upkeep of radio equipment to ensure optimal performance.

- **Mayday**: A term indicating a life-threatening emergency.

- **Negative**: No.

- **Out**: Indicates that the speaker has finished talking and no reply is expected.

- **Over**: Indicates that the speaker has finished talking and is awaiting a reply.

- **Phonetic Alphabet**: Used to spell out words clearly over the radio, e.g., A - Alpha, B - Bravo.

- **Programming**: The process of pre-setting important frequencies into the radio.

- **Protocol Understanding**: Knowledge of the specific systems and standards used in communication.

- **Reception**: The initial stage where radio signals are captured by an antenna or receiver.

- **Say Again**: A request to repeat the last message.

- **Stand by**: Wait.

- **Voice Scramble**: A feature that provides basic security by altering the sound of voice transmissions.

Made in the USA
Monee, IL
15 February 2024

53565831R00083